SINGAPORE PERSPECTIVES 2007

A New Singapore

SINGAPORE PERSPECTIVES 2007

A New Singapore

Edited by
Tan Tarn How

Published by

World Scientific Publishing Co. Pte. Ltd.
5 Toh Tuck Link, Singapore 596224
USA office: 27 Warren Street, Suite 401-402, Hackensack, NJ 07601
UK office: 57 Shelton Street, Covent Garden, London WC2H 9HE

British Library Cataloguing-in-Publication Data
A catalogue record for this book is available from the British Library.

SINGAPORE PERSPECTIVES 2007: A NEW SINGAPORE

ISBN-13 978-981-270-571-6 (pbk)
ISBN-10 981-270-571-6 (pbk)

Printed in Singapore.

Contents

Preface

The conference theme for Singapore Perspectives 2007 was "A New Singapore". It was one, when I started to conceive the programme, that I thought would be particularly apt at a time when the nation appeared to be on the cusp of changes that would likely prove to be momentous. I wished the conference to discuss some of these societal transformations — whether impending or already ongoing — and to perhaps suggest solutions to potential problems.

The tradition of our annual Perspectives conference is to have a panel on economics (the others being politics, society and foreign affairs), and in this area, one of the most interesting questions of late is the disparate effects of globalisation on different segments of society, the poor and the rich, the domestic and the globalised. Chua Hak Bin and Manu Bhaskaran's papers deal with that, including the politically delicate issue of addressing the income-suppressing consequences on the poor of an aggressively open economy and an open-door policy towards foreigners. Both ask this fundamental question: who is the growth of the economy for — the transient workers, the Permanent Residents or Singaporeans? Both, as is also Perspectives tradition, gaze into their expert crystal balls to glean the prospects for the economy. As can be seen from their papers, they more than adequately acquit themselves.

The second panel grapples with the "New Politics of Identity" in the face of immigration policies and the Singapore diaspora. Brenda S. A. Yeoh outlines the changing demographic make-up of Singapore, namely, the increasing diversity that has arisen through the arrival of "non-traditional sources" of residents and citizens, many as marriage partners. Greater efforts need to be made to integrate not just these people but also the "use-and-discard" foreign workers that come here as maids and low-end wage labour. This call is echoed by Kwok Kian Woon, whose wide-ranging and philosophical reflections on identity are a reminder of the immensely complex nature of the issue. Interestingly, his discussion ends with the notion of the "cultural clash" caused by diversity, positing that it is productive of creativity, and questioning whether Singapore's predilection for safety and security is therefore counterproductive to the flowering of new thinking. Beatrice Chia-Richmond brings her own humourous and endearingly off-beat take on not just her identity but the state of the arts in Singapore. If there is a paper in this volume that needs to be heard rather than read for its fullest effect, then the director/actress' is the best example. Harming by charming is her subtle method: for instance, her criticism of censorship in the infamous Josef Ng snipping of pubic hair incident at the end of her speech is made the more cutting by the counterpoint providing at the beginning when she professes unwillingness to engage in "whining" about the arts scene and the establishment.

In the third panel on the impact of the Internet on society, Cherian George discusses the differences — and the connection — between traditional, controlled broadcast/print media and unbridled online media. His view is that the latter has led to the formation of a public sphere — a network for communicating information and points of view — that is informal and works on different rules from the formal public sphere of old media. His paper segues nicely into the second by Lim Sun Sun, whose main message is the need to ramp up the hitherto insufficient efforts at equipping Internet users with the deconstructive tools of media literacy. She also takes the subject back to questions raised by George, venturing that the unwillingness to narrow the gap between functional and critical media literacy has causes that are not so much educational as political, namely, the desire of the government to control traditional and now online media.

The last session dwells not so much on a new Singapore as the possibility of "A New ASEAN". The implications for Singapore of whether and how the regional grouping reinvents itself may not be made explicit in the presentations by the three speakers but there are no doubts that they will be far-reaching. Azmi Mat Akhir's paper, delivered on behalf of the ASEAN Secretariat, deals with why change is imperative. Amitav Acharya provides a perceptive historical account of the association's successes and analyses the tightrope that it must walk in balancing opposing demands as it remakes itself. Rodolfo Severino paints a stark picture of the three possibilities for the charter that ASEAN is drafting, and warns that it is actually better to have a weak charter than a strong one that its member states do not comply with.

I wish to thank Claris Wang for being a first-rate administrator of the conference and her gentle persistence in pushing me to the necessary and to do it on time. I also wish to thank Yeoh Lam Keong and Manu Bhaskaran for helping to conceptualise the economics panel, and Gillian Koh for putting together the panel on identity. Thanks also to the copy-editor Pang Gek Choo; Cheong Chean Chian of our publisher World Scientific Publishing; and transcribers Andrea Wee, Luu Tran Huynh Loan, Grace Cheow and David Ho.

I also wish to give my special thanks to my research officer Rica Agnes Castaneda for handling the myriad duties involved in getting this book out.

Tan Tarn How

Acknowledgements

Corporate Sponsors

IPS is grateful to the following institutions for their support of Singapore Perspectives 2007 held on Thursday, 11 January 2007.

Principal Sponsors

TEMASEK

HOLDINGS

Sponsors

ASIA PACIFIC
ASIA PACIFIC BREWERIES LIMITED

HSBC
The world's local bank

HOUSING &
DEVELOPMENT
BOARD

MPA
SINGAPORE

NYP
Nanyang
Polytechnic

NANYANG
TECHNOLOGICAL
UNIVERSITY

NUS
National University
of Singapore

NGEE ANN
POLYTECHNIC

osim
Inspiring Life

PHILIPS

SMU
SINGAPORE MANAGEMENT
UNIVERSITY

SINGAPORE
POLYTECHNIC

SMRT

TEMASEK
POLYTECHNIC

CHAPTER 1

Chairman's Introductory Remarks

TOMMY KOH

Your Excellencies, Governors, Sponsors, Corporate Associates and Friends of IPS, a very warm welcome to Perspectives 2007.

The response to this year's conference has been overwhelming. We have a record number of 750 participants. Irene Lim, our Administration Manager, has had to turn away over 100 requests because of the lack of space. If the number keeps growing each year, we may have to consider moving to a bigger venue. I thank you for your support.

I would also like to thank our two principal sponsors, Temasek Holdings and the Standard Chartered Bank, as well as the other 15 sponsors, which are Asia Pacific Breweries, Housing & Development Board, Maritime and Port Authority of Singapore, Nanyang Polytechnic, Nanyang Technological University, National University of Singapore, Ngee Ann Polytechnic, OSIM International, Philips Electronics Singapore, Shell Companies in Singapore, Singapore Management University, Singapore Polytechnic, SMRT Corporation Ltd, Temasek Polytechnic and the Hongkong and Shanghai Banking Corporation. I congratulate Chang Li Lin, our Public Affairs Manager, for succeeding in increasing the number of sponsors each year.

It is the custom in IPS to have a different colleague from the research wing of the family and one from the administrative wing to take the lead in organising the annual Perspectives conference. This year, the two leaders are Senior Research Fellow Tan Tarn How and Administration Officer Claris Wang, and I wish to thank them for their outstanding jobs. I would also like to thank Acting Director Arun Mahizhnan and Irene Lim for overseeing the whole process.

The theme of this year's Perspectives is "A New Singapore". The conference will address four new trends and developments:

(i) How will the large influx of new immigrants and the increasing outflow of Singaporeans affect our social and political compact and our nation-building project?
(ii) How will the new technologies, such as the Internet and mobile phone, affect our civil society, our media and our political governance
(iii) How will the trend towards a dual economy — a robust globalised economy co-existing with a sluggish domestic economy — and the widening social divide, affect our social cohesion and harmony?
(iv) Will ASEAN, which will be 40 years old in August, succeed in seeking to re-invent itself by adopting a legal personality, a charter, embracing deeper integration and strengthening its institutions and effectiveness?

The Singapore government has appointed me to represent Singapore in the High-level Task Force to draft the ASEAN Charter. I will be leaving Singapore tomorrow morning for Cebu, the Philippines, to attend a joint meeting between the Task Force and the ASEAN Charter's Eminent Persons Group (EPG). Our Deputy Prime Minister, Professor S. Jayakumar, represents Singapore in the EPG. I look forward to this endeavour and hope that my colleagues and I will succeed in drafting a good charter for adoption by the ASEAN Summit, when it convenes in Singapore in November this year.

Let me conclude by summarising the Institute's main achievements in 2006:

(i) It co-organised, with the International Monetary Fund and the World Bank, the highly successful 2006 Programme of Seminars;
(ii) It commissioned a survey on the 2006 General Elections and organised a Post-Election Forum;

(iii) It launched with the Nanyang Technological University the inaugural annual index ranking the 10 ASEAN economies, 35 states and union territories of India, 31 provinces of China, Hong Kong, Macau and Taiwan;

(iv) It organised the 5th Young Singaporeans Conference;

(v) It organised three important public lectures for His Eminence Cardinal Renato Raffaele Martino, President of the Pontifical Council for Justice and Peace and Special Envoy of His Holiness Pope Benedict XVI; Professor Larry Diamond of the Hoover Institution, Stanford University; and Professor Rosabeth Moss Kanter of the Harvard Business School;

(vi) It co-published two books with the World Bank, one on the impact of China and India on the world economy and the other containing essays on Asia's future by 17 Asian thinkers; published an important new book on madrasah education in Singapore and five other publications;

(vii) It continued to co-organise, with The Business Times, the Singapore Economic Roundtable twice a year; and

(viii) It organised 16 breakfast meetings, working lunches and dinners and other events for our 97 Corporate Associates. Our key performance indicator for 2007 is 110 members.

I will now request my colleague, Mr Yeoh Lam Keong, the Chairman of the first panel, to take over the proceedings.

Thank you.

SECTION 1

The New and Dual Economy

CHAPTER 2

Singapore Economy: The New and The Dual

CHUA HAK BIN

There appears to be two faces to the current economic expansion: a new and a dual. The *new economy* thesis argues that Singapore is undergoing some form of Renaissance, with sustained and higher growth than the typical speed limit likely over the medium term. The *dual economy* thesis highlights that despite rosy headline growth, there appears to be divergent growth patterns persisting between different businesses, income groups and even within certain asset classes, particularly residential property (Chua and Lim, 2006). The two features need not necessarily be inconsistent, and on the contrary, may be part and parcel of globalisation.

THE NEW ECONOMY

The conviction about a *new economy* comes from robust economic expansion over the last few years, since the Sars crisis, with GDP growth averaging 7.6 per cent over 2004–2006 (see Chart 1). Growth going forward will likely come in above the 3–5 per cent range, previously regarded as the long-term growth speed limit for a mature economy. The government has reiterated its confidence by forecasting a GDP growth range of 4–6 per cent in 2007, a departure from the last three years when the initial official forecast started at 3–5 per cent.

Chart 1 GDP growth trending up since 2003 Sars crisis: Renaissance or luck?

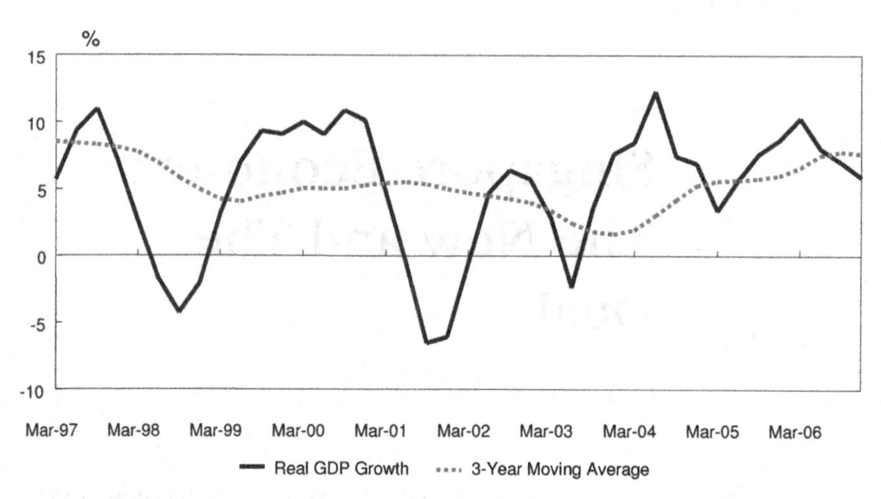

SOURCE CEIC Data Company Limited.

Luck may have certainly been a factor, given the relatively benign global environment over the last few years since 2003. But there are nevertheless structural developments supporting the Renaissance thesis. The relative out-performance of the Singapore economy against its regional peers suggests more fundamental forces at work, some of which are policy-induced. We believe the new economy is a result of (1) enhanced competitiveness arising from an aggressive tax and Central Provident Fund (CPF) restructuring; (2) a more liberal immigration policy; (3) a more diversified economy with new growth drivers, and (4) a pro-growth approach.

Enhanced Competitiveness from Tax and CPF Restructuring

Tax and CPF restructuring may have increased economic competitiveness, unleashing new growth drivers and wider domestic investment opportunities. Aggressive tax and CPF cuts from 2001 have significantly reduced costs for employers. Corporate tax rates have been cut from 26 per cent to 20 per cent over 2001–2006 (see Table 1). Prime Minister Lee Hsien Loong has indicated that, alongside a further Goods and

Services Tax (GST) increase, more income tax cuts could be possible going forward (Lee, 13 November 2006). Employer CPF contribution has also been cut, with the contribution rate cut by 3 percentage points to 13 per cent from 16 per cent and the salary ceiling gradually reduced from S$6,000 to S$4,500 over three years (see Table 2).

Table 1 The grand tax restructuring: More to come?

Year of Assessment	Corporate Tax Rate	Personal Tax Rate	GST
1997–2000	26%	Top Rate 28%	3% (since 1 April 1994)
2001	25.5%	Top Rate 28%	3%
2002	24.5%	Top Rate 26%	3%
2003	22%	Top Rate 22%	4%
2004	22%	Top Rate 22%	5%
2005	20%	Top Rate 22%	5%
2006	20%	Top Rate 21%	5%
2007	Towards 18%?	Top Rate 20%	Towards 7%

SOURCE www.iras.gov.sg; Citigroup estimates. See Chua, 20 November 2006.

Table 2 CPF restructuring 2003–2006

	Major CPF Policy Changes to Employer Contributions
2003	On 1 October, employer CPF contribution rate cut by 3 percentage points to 13% from 16% previously (for workers above 55)
2004	On 1 January, CPF salary ceiling was lowered to $5,500 from $6,000
2005	On 1 January, CPF salary ceiling was lowered to $5,000 from $5,500; employer CPF contribution rate for older workers aged 50–55 was also reduced from 11% to 9%
2006	On 1 January, CPF salary ceiling was lowered to $4,500 from $5,000; employer CPF contribution rate for older workers aged 50–55 was reduced from 9% to 7%

SOURCE www.cpf.gov.sg. See Chua, August 2006.

Verdict on the economic restructuring is probably best reflected in the strong job growth figures and rising foreign investment over the last few years. Job creation has gradually risen from a contraction of 12,950 in 2003 to 71,400 in 2004 and 113,300 in 2005. For the first nine months of 2006, job growth at 123,000 has already beaten the whole of 2005. Overall unemployment rate has slid back from a high of 4.8 per cent in September 2003 to 2.7 per cent in September 2006.

Such strong job growth figures are spectacular and may not have been possible without the generous tax and CPF cuts, which helped to lift corporate profitability and attract fresh investments. Net manufacturing investment commitments recovered from S$7.5 billion in 2003 to S$8.3 billion in 2004 and S$8.5 billion in 2005. For the first nine months of 2006, net manufacturing investment commitments are holding up at about the same levels, of about S$6 billion. Figures on the number of newly registered companies are probably more representative, as they include services. Newly registered companies have risen to 19,501 in 2005 and 17,153 in 2004 from the low point of 13,544 in 2003. The figure for 2006 should come in slightly above the 2005 level and almost double the 2000–2002 levels. About 90 per cent of the new companies are in services.

More Liberal Immigration Policy

A liberal immigration policy, even more relaxed than during the early 1990s boom, is increasing the influx of foreign talent, making higher potential growth possible. Permanent residents, for example, grew 8.7 per cent in 2000–2005, ten times the growth rate of citizens. The government is studying its population policies and more relaxation may be in store to increase the citizenship take-up rate. The number of new citizenships granted in 2005, at 12,900, is almost double the rate of previous years (see Table 3). Preliminary figures for 2006 and government policy direction suggest that this trend will continue.

The more relaxed immigration policy has made the current economic boom possible, as a large fraction of the strong job demand is satisfied by foreign workers. Labour force growth, currently running at 6 per cent, is significantly higher and about double the pace even compared to the boom during the early 1990s (Chua, 7 November 2006). About half of the new jobs created are going to non-residents.

Table 3 Demographic trends, 1990–2005

	Number ('000)			Per Cent of Population			Average Annual Growth (%)		
	2005	**2000**	**1990**	**2005**	**2000**	**1990**	**1990–2005**	**2000–2005**	**1990–2000**
Resident Population	3,554	3,263	2,736	81.7	81.2	89.8	1.8	1.7	1.8
Citizens	3,113	2,973	2,623	71.5	74.0	86.1	1.1	0.9	1.3
Permanent Residents	441	290	112	10.1	7.2	3.7	9.6	8.7	10.0
Non-Resident Population	798	755	311	18.3	18.8	10.2	6.5	1.1	9.3
Total Population	**4,351**	**4,018**	**3,047**	**100.0**	**100.0**	**100.0**	**2.4**	**1.6**	**2.8**
Foreign Population	1,239	1,045	423	28.5	26.0	13.9	7.4	3.5	9.5

SOURCE Singapore Department of Statistics, Demographic Trends, *General Household Survey 2005*; Koh *et al.*, 2002; Citigroup estimates. See Chua, September 2006.

NOTE Full time series for permanent residents and citizens are not available.

Table 4 New citizens in Singapore: Sharp jump in 2005 and 2006

Year	Number of Citizenships Granted	Fertility Rate
2001	6,500	1.41
2002	7,600	1.37
2003	6,800	1.25
2004	7,600	1.24
2005	12,900	1.24
2006 Jan–June	6,800	. . .

SOURCE *New Straits Times*, 24 August 2006; Singapore Department of Statistics, *Population Trends 2005*.

Chart 2 Current boom seeing strong increases in jobs and foreign workers

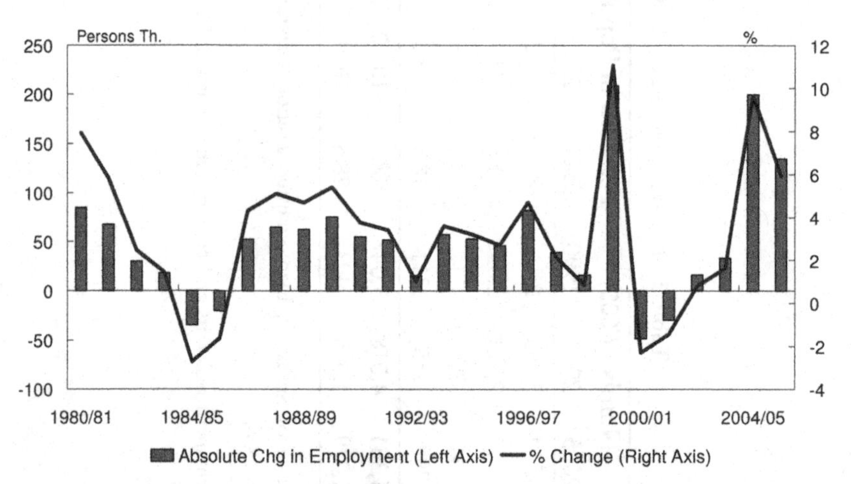

SOURCE CEIC Data Company Limited.

The government will have to balance overly-aggressive immigration targets against concerns about job security and low wage growth among the lower-income resident households.[1] Liberalisation will therefore likely be gradual and calibrated. A successful population policy will have a material impact on growth, property and businesses where domestic critical mass matters. It will also stall the erosion on growth coming from a low fertility rate and an ageing population. Greater differentiation in the policy treatment of citizens, permanent residents and foreigners is also expected, which will increase the incentive for taking up citizenship.

More Diversified Economy and New Growth Drivers

Globalisation and government policies have created new growth drivers, particularly in the areas of financial services (private banking and wealth management), biomedical (pharmaceuticals), luxury-end property and tourism (integrated resorts). Multiple growth drivers increase the likelihood of sustained growth and reduce the risks from any single-engine failure. Diversification from electronics, in particular, will reduce the vulnerability to volatile tech cycles. Electronics as a share of GDP has fallen to about

Table 5 Diversification from electronics

Percentage of GDP	2005	2000	1995
Goods Producing Industries	**31.1**	**33.1**	**32.6**
Manufacturing	26.1	25.8	25.2
Electronics	9.4	12.3	12.0
Biomedical	4.8	...	...
Chemical	3.7	4.1	2.4
Transport Engineering	2.6	1.3	1.6
Construction	3.4	5.6	5.7
Utilities	1.5	1.6	1.5
Services Producing Industries	**63.1**	**61.2**	**61.5**
Wholesale & Retail Trade	15.8	12.8	14.3
Business Services	12.6	13.7	13.3
Transport & Communications	11.8	11.5	10.8
Financial Services	10.7	10.8	11.0
Hotels & Restaurants	1.8	2.1	2.4
Other Service Industries	10.4	10.2	9.5

SOURCE CEIC Data Company Limited.

NOTE Percentages may not add up to 100 per cent due to need to include ownership of dwellings (imputed) and exclude financial intermediation services indirectly measured.

9.4 per cent in 2005 from about 12.3 per cent in 2000, with biomedical increasing its share to about 4.8 per cent of GDP (see Table 5). Services have also steadily risen to about 63.1 per cent of GDP in 2005 from about 61.2 per cent in 2000.

Manufacturing has become noticeably less correlated with the tech cycle in recent years with the growing importance of the biomedical sector (see Chart 3). The correlation between industrial and electronics output was, for example, a tight 0.86 in the period 2000–2002. That correlation has dropped to only 0.36 in 2004-2006.[2] Manufacturing, for example, in the current tech slowdown, has been holding up relatively well because of strong pharmaceutical growth. Diversification away from electronics may mean that the Singapore economy will be less vulnerable to volatile tech swings in the future.

Chart 3 Manufacturing becoming less correlated with tech cycle in recent years

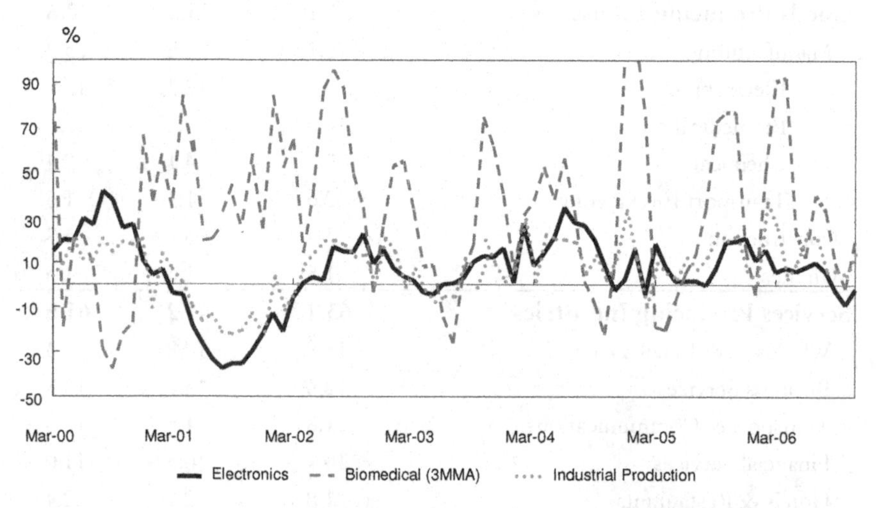

SOURCE CEIC Data Company Limited.

Services will also play an increasingly important role in diversifying and driving the economy going forward. Tourism, in particular, will be a key growth driver, particularly from 2010. The two integrated resorts — Marina Bay Sands and Resorts World at Sentosa — will help double tourist arrivals to 17 million and triple tourism receipts to S$30 billion by 2015. These targets are well within reach and may well be exceeded, judging by Macau's experience (see Chart 4). Universal Studios alone (part of Resorts World) should attract five million visitors annually; while Resorts World itself could achieve 10 million. On top of this is the draw of Marina Bay Sands, which is targeting the lucrative conventions and exhibitions market. Tourism's contribution to GDP has declined from about 6.1 per cent of GDP in 1993 to 3 per cent in 2002, according to government estimates. But the sector should turn around smartly, with tourism climbing back up to about 5 per cent of GDP in 2015, by our estimates (Chua, December 2006).

Chart 4 Tourist arrivals in Macau now double that of Singapore post-2001 liberalisation

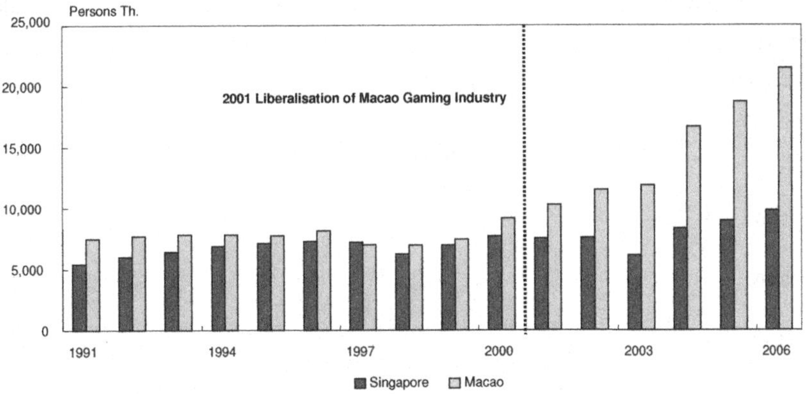

SOURCE DSEC, CEIC Data Company Limited, Citigroup estimates. See Chua, December 2006.

New Philosophy: Growing "As Fast As We Can"

PM Lee Hsien Loong, in his National Day Rally Speech on 20 August 2006, said:

> "I think that when conditions are good and the sun is shining, we should go for it, as fast as we can, as much as we can. Get the growth, put it under our belt, put it aside a little bit, so when the thunderstorm comes again, we will be ready."

We interpret the growing "as fast as we can" philosophy as being more accommodative in terms of policy and key resources, particularly towards foreign labour, land and investment projects (Chua, 13 November 2006). This will likely imply higher GDP growth over the boom cycle. The shift towards an aggressive pro-growth attitude may be because of the economic volatility experienced since 2000, with two sharp downturns, in 2001 (tech bust) and 2003 (Sars crisis). The economy appeared to have borne the brunt of the downswings when external conditions turned ugly — but may not have capitalised on the good times as strongly when conditions improved.

The government appears to be more willing to capitalise on the current boom, being more aggressive in releasing land sites and locking in investment projects, which will help cushion the impact from any sudden downturn. Commercial land sites, for example, have soared after a long dry spell of seven years (see Chart 5). As such projects and investments take several years to materialise, growth from such commitments will be secured over several years, cushioning the impact from any subsequent downturn.

Chart 5 Soaring commercial land sales after long dry spell

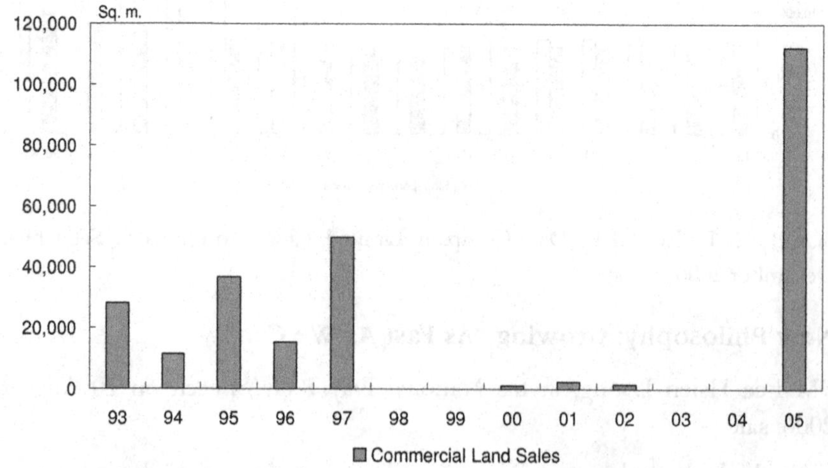

Sq. m.
120,000
100,000
80,000
60,000
40,000
20,000
0
93 94 95 96 97 98 99 00 01 02 03 04 05
Commercial Land Sales

SOURCE CEIC Data Company Limited, Urban Development Authority.

THE DUAL ECONOMY

The *dual economy* thesis highlights that despite rosy headline growth, there appears to be two parts to the economy, moving along different frequencies (Chua and Lim, 2006). Businesses catering to the global market are witnessing more robust growth, while those serving the domestic market are seeing more sluggish growth. Divergent growth patterns between different income groups, businesses and even within certain asset classes, particularly property, have become quite stark. Globalisation and policies on tax, CPF and immigration — while necessary to increase competitiveness — may have also accentuated these divergent forces.

External demand has been the key growth driver over the last few years, with domestic demand remaining relatively sluggish. Exports as a share of GDP has surged as a result (see Chart 6). This suggests that the changes could be more structural — driven by policies capitalising on globalisation — than just cyclical in nature. This contrasts quite noticeably with the early-1990s boom when domestic demand played a much more significant role. The sharp rise in exports has, in particular, been driven by rapid growth in new drivers such as pharmaceuticals and offshore and marine transport.

Private consumption is growing at less than 3 per cent despite headline GDP growth of about 8 per cent (see Chart 7). Wages across different income segments are diverging: the recent household survey showed that the bottom 30 per cent income percentile saw incomes fall over the period 2000–2005. Wages of the highest-income segment in contrast saw large gains (see Table 6). Owners of capital are seeing a windfall, while workers are seeing nominal wages barely keeping up with inflation.

The residential property market has been seeing a boom in the luxury end, but the mass residential market remains soft (Chart 9). The banking sector has been seeing a boom in offshore lending and private banking, but domestic SGD lending, particularly consumer lending, remains sluggish (Chart 8). Consumer loan growth is growing at about 1.5 per cent in October, with mortgages expanding at 2.1 per cent. Construction, despite emerging from a seven-year drought, is barely growing, versus a boom in marine and aviation transport, reflecting strong regional opportunities.

Chart 6 Surge in exports as % of GDP over last 5 years

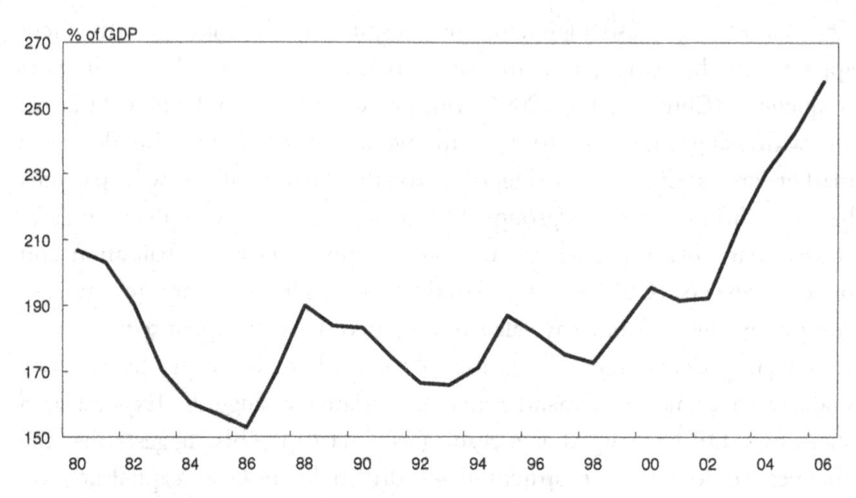

SOURCE CEIC Data Company Limited, Citigroup estimates.

Chart 7 Consumption remarkably sluggish despite rosy headline GDP growth

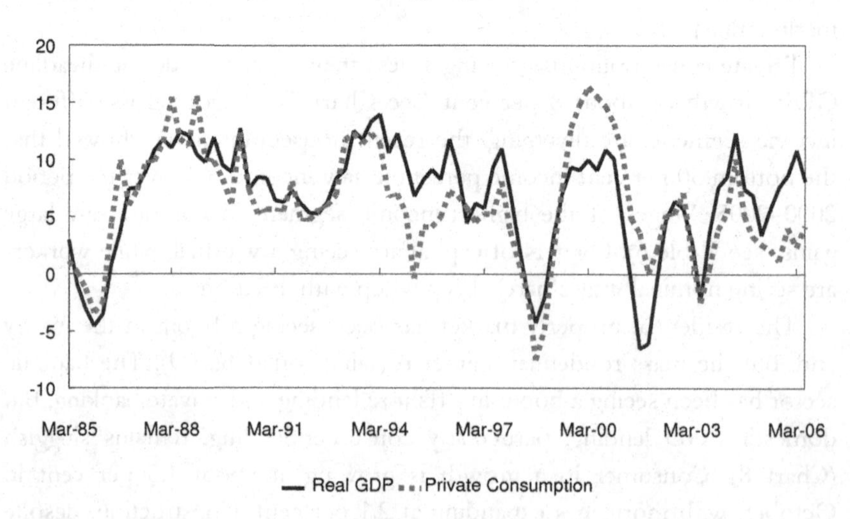

SOURCE CEIC Data Company Limited, Citigroup estimates.

Chart 8 Divergence between offshore ACU and domestic SGD lending (% y/y change)

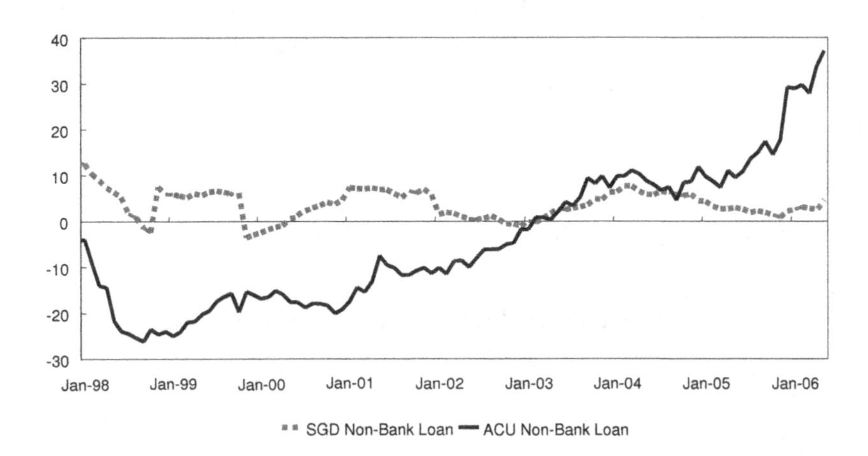

SOURCE CEIC Data Company Limited, Citigroup estimates.

Chart 9 Diverging property price indices across regions, with strong price gains in central region

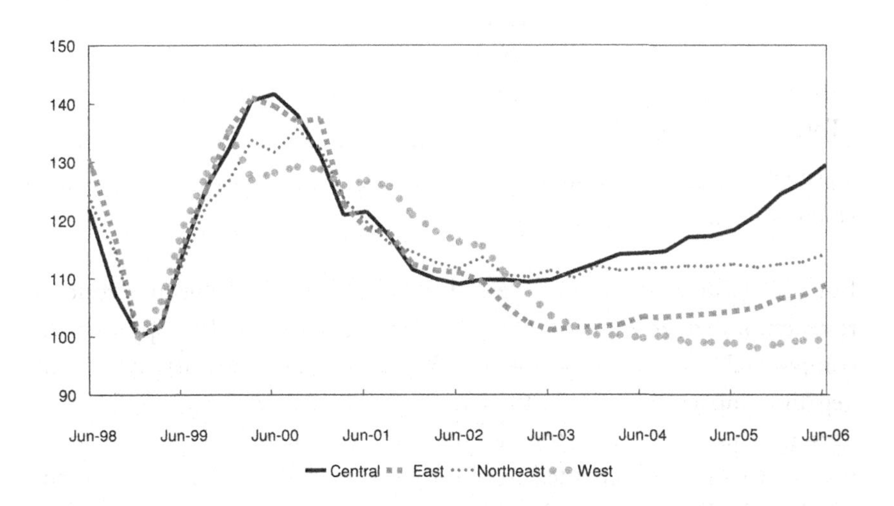

SOURCE CEIC Data Company Limited, Citigroup estimates.

Embracing globalisation has disproportionate effects on different income groups, particularly in a small open economy where the impact is amplified. Such trends favour the higher-income and foreign segment groups – those in the "global periphery" — who are benefiting from the regional boom. The lower-income groups have to face greater competition because of globalisation and liberalisation, particularly with the emergence and opening up of China and India.

Table 6 Average monthly household income from work by decile among all resident households

	Average Household Income (S$)			Average Annual Chg (%)
	2000	2004	2005	
1st–10th	90	…	…	…
11th–20th	1,470	1,170	1,180	–4.3
21st–30th	2,250	2,140	2,190	–0.5
31st–40th	2.950	2,890	2,990	0.3
41st–50th	3,660	3,670	3,850	1.0
51st–60th	4,470	4,600	4,840	1.6
61st–70th	5,390	5,510	5,890	1.8
71st–80th	6,520	6,820	7,260	2.2
81st–90th	8,270	8,960	9,300	2.4
91st–100th	14,360	15,960	16,480	2.8
Total	**4,940**	**5,170**	**5,400**	**1.8**

SOURCE Singapore Department of Statistics, Key Findings of the General Household Survey 2005, June 2006.

But globalisation may be only part of the reason for the dualism. Economic restructuring over the last few years may have also disproportionately compressed the wages of the middle class (Chua, August 2006). The negative impact from the CPF cuts has outweighed the gains from the personal income tax cuts for this segment. The high-income segment, on the other hand, has benefited from the restructuring as the income tax cuts dominated. The middle-class squeeze may help partly explain the emergence of a dual economy.

The middle-class group has been squeezed by the CPF cuts, with the impact outweighing the impact from tax cuts (see Chart 10). We estimate the cumulative impact on average wages from the CPF cuts over 2003–2006 to be negative 3.6 per cent and from income tax cuts over 2002–2007 to be positive 0.8 per cent. The overall net impact on average wages, is therefore, negative 2.8 per cent, a non-trivial sum. The impact profile over the adjustment period and across income groups was moreover uneven.

Chart 10 An upper middle class squeeze (cumulative % impact on wages from both CPF and tax changes across income decile groups, 2001–2007)

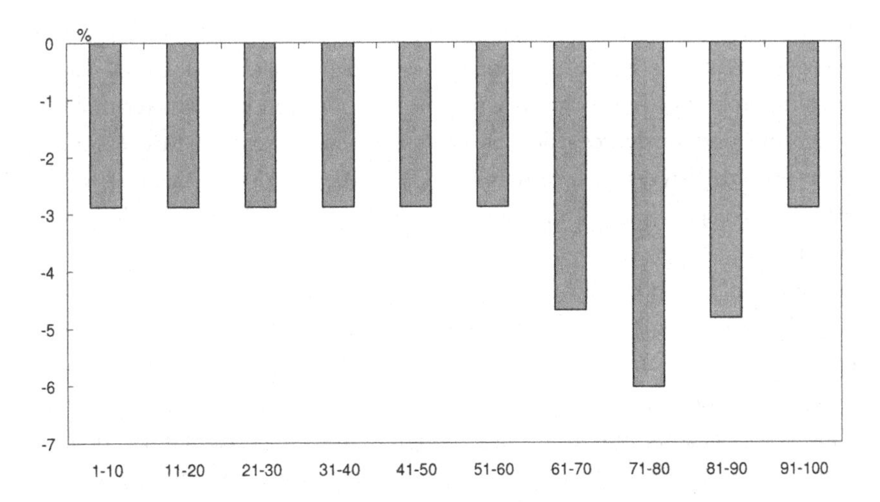

SOURCE Citigroup estimates. See Chua, August 2006.

The negative impact was largest on the middle class. The upper middle class squeeze is largely because of the greater negative impact from lower CPF salary ceiling limits; relatively larger fall in the CPF tax relief; and smaller percentage income gains from tax cuts relative to the top bracket segment. This upper middle class segment represents HDB upgraders and may help explain the sluggish upgrading demand for mass residential private property.

The proposed CPF increase of 1–2 percentage points in 2007 will help to *partly* negate the brunt of this aggressive restructuring and provide relief to the middle-class segment. Fiscal transfers in the past have not really cushioned the impact on the middle class from globalisation and restructuring, as these measures have been directed at mainly the lower income. A Monetary Authority of Singapore study estimates that special transfers in the FY06 budget would increase real consumption growth by 0.48 percentage points for one- to four-room HDB households, 0.11 percentage points for five-room, executive and HUDC households, and 0.02 percentage points for private households (Monetary Authority of Singapore, 2006: 66).

CONCLUSION

Embracing globalisation has generated huge economic gains for Singapore. Policies favouring lower income taxes, foreign talent and global capital have made higher potential growth possible over the medium term, above what was previously regarded as the speed limit for a mature economy. New growth drivers have emerged, which tap on rising Asian affluence, intra-Asian and global production networks, and a mobile global talent pool.

But globalisation brings with it fresh challenges. That pressure is amplified in a small open economy where the impact has been most acute on the lower income group. Several other Asian governments have not chosen the more liberal strategy precisely because of such repercussions, with some closing the door on foreign workers or capital for fear of competition and pressure on the local population. Such protectionist policies, unfortunately, only hurt growth in the longer term and slow down innovation and change. Singapore has chosen to deal with the emerging widening income gaps with appropriate fiscal policies, education and an enhanced social safety net. So far, the results on growth are clearly showing, with a fiscal windfall to cope with the pressures from globalisation.

ENDNOTES

1. For a fuller discussion, see Chua, September 2006.

2. 2006 figures only up to November, the latest available month.

REFERENCES

Chua, H. B. "Growing 'As Fast As We Can'," *Citigroup Singapore Market Weekly*, 13 November 2006.

Chua, H. B. "Integrated Resorts: Will Singapore See A Macau-Type Boom?" *Citigroup Singapore Market Weekly*, 11 December 2006.

Chua, H. B. and Lim J. S. "Singapore: A Dual Economy?" *Citigroup Asia Macro Views*, 25 July 2006.

Chua, H. B. "Singapore: Comparing the Current Boom to the Early 1990s — Sustainable or Bust in the Making," *Citigroup Asia Macro Views*, 7 November 2006.

Chua, H. B. "Singapore's Population Policy: A Matter of Life and Death," *Citigroup Asia Macro Views,* 21 September 2006.

Chua, H. B. "Singapore — Restructuring and an Upper Middle Class Squeeze", *Citigroup Asia Macro Views*, 4 August 2006.

Chua, H. B. "Tax Restructuring: Chapter 2?" *Citigroup Singapore Market Weekly*, 20 November 2006.

Koh, A. T. *et al. Singapore Economy in the 21st Century: Issues and Strategies.* McGraw-Hill, Singapore, 2002.

Lee, Hsien Loong. Speech in Parliament, 13 November 2006.

Monetary Authority of Singapore, "Table 3.13: Impact of Selected Special Transfers in FY2006," *Macroeconomic Review*, April 2006.

CHAPTER 3

Singapore Economy: Medium-Term Outlook

MANU BHASKARAN

INTRODUCTION

I would like to focus on two main areas in my survey of the prospects for the Singapore economy. First, I will look at the immediate-term prospects for the economy by examining the cyclical forces that are at work and to then assess the resilience of Singapore's economy. Second, I will discuss some of the drivers of longer-term trends in the economy. In this section, what really crystallises is a key question: "Who is all this growth for?" The point I am making is even though we may be able to report very high GDP growth rates in future, the question of who really benefits from that growth will become an increasingly important political and policy question.

New Phase of Growth

Let us just take a step back and look at Singapore's economic growth in the last 15 years, as shown in Chart 1. It has certainly been an eventful period. We went through a period of very high growth from the late 1980s that effectively ended with the Asian financial crisis in 1997. A recovery period followed the crisis but unfortunately, this recovery was interrupted by a series of shocks — Sars, 9/11, terrorism — never-ending shocks, it seemed, at one point. This was also a period when the economy and businesses re-engineered in response to a more competitive environment. This paper will argue that this period of unusual difficulties is now over and a more resilient and vibrant Singapore economy is now in a new phase of growth.

Chart 1 A new phase of growth

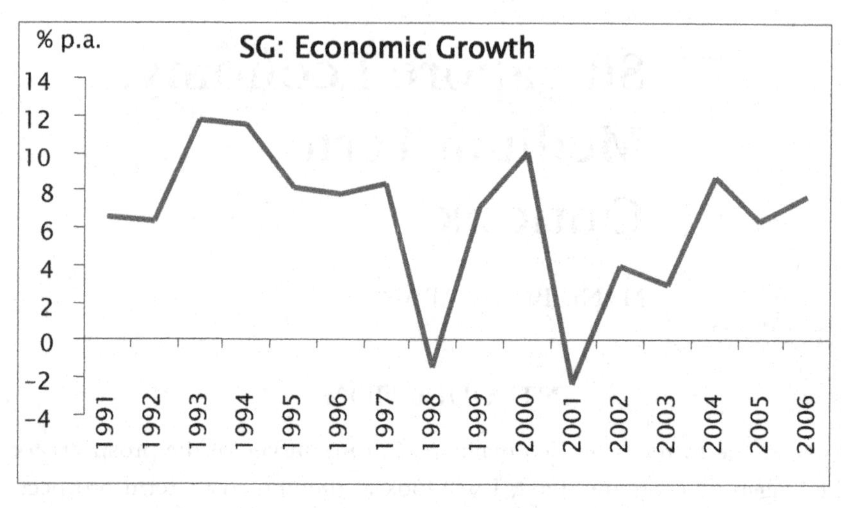

SOURCE Collated by Centennial Group using CEIC Database.

Near-Term Prospects May Include Some Shocks

The lead indicators point to a rising risk of a moderate cyclical downturn in the external demand for our goods and services. However, that is the baseline scenario, but the trouble is that this baseline scenario is subject to a much greater margin of error than previously because there are so many pressures building up in geopolitics as well as in terms of the structural weaknesses in the global economy that the risks of huge deviations from our baseline are quite high. In short, Singapore must be ready for unanticipated shocks. It could be a financial accident in one of the large economies that triggers off a loss of confidence. It could be a disruption in the global flows of financial liquidity. Perhaps the yen-carry trade gets disrupted and then we get a mass exodus of funds from emerging markets. It is hard to know what exactly will cause the shock but the chances are that, over the next one to two years, Singapore will have to deal with the consequences of a few such shocks.

But what is helpful for Singapore is that the cyclical problems can be mitigated by a number of factors particularly in terms of how technology demand, which is still very important for Singapore, is holding firm. As far as the US economy is concerned, it is the nature of the economic slowdown there that matters rather than just the fact of the slowdown. If the US slowdown is driven by slower growth in consumer spending there, with no recession and with rising business spending, then Singapore can still do well. But if both business spending and consumer spending in the US go down, then we are in trouble.

Cyclical Risks Rising in Global Demand

Let me expand on these issues. In trying to look at the near-term, the best lead indicator that I have found is the Organisation for Economic Co-operation and Development's (OECD's) lead indicators for the major economies. As you can see from Chart 2, the picture is very clear. There is going to be a global economic deceleration despite the fact that everyone feels extremely good right now. It will be inevitable. We have to be prepared for it for the next six months, but I think it will be contained.

Chart 2 OECD lead indicators — Risk for external demand?

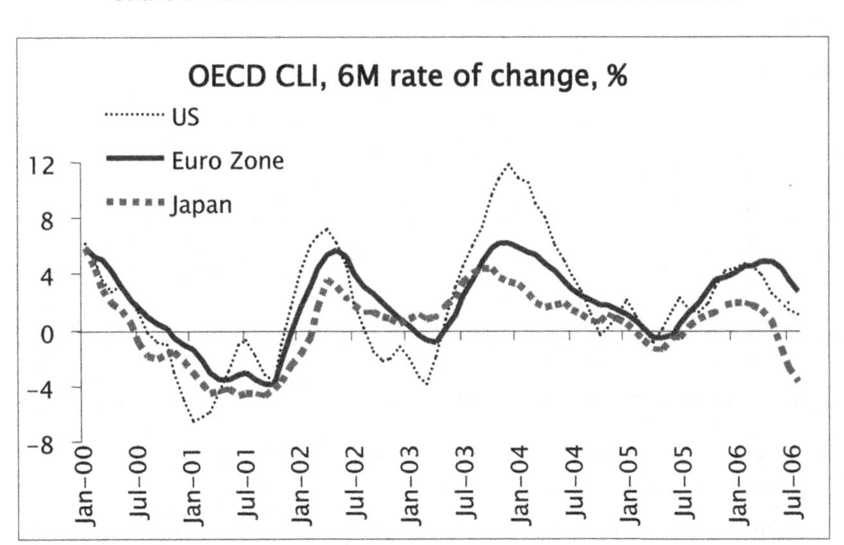

SOURCE Collated by Centennial Group using OECD Database.

Electronics Less But Still Important

The correlation of Singapore's industrial production with electronics has diminished significantly over the last few years. However, the local economy is still very much influenced by electronics, which still makes up about 36 per cent of industrial production. Moreover, it is a growing segment of our non-oil re-export trade. This contributes materially to the wholesale/retail sector that forms 16 per cent of the economy. Electronics does influence many other parts of the economy such as air cargo as well. If the demand for computers and electronic products can hold up (see Chart 3), that will help provide some offsets to the overall global economic cycle.

Chart 3 Mitigating factor — Tech demand?

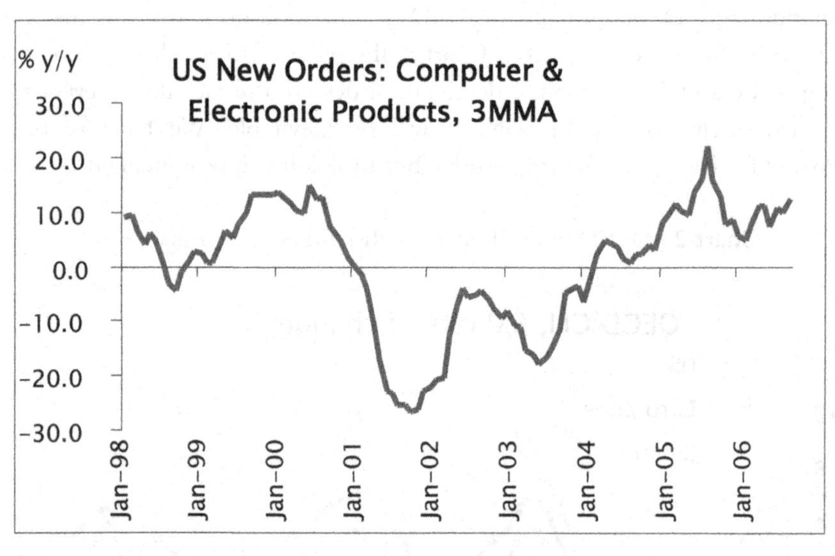

SOURCE Collated by Centennial Group using CEIC Database.

Regional Indicators Look Good

The other key driver that is very important for Singapore is the region, and fortunately, there are reasonably good lead indicators for many of our key economies (see Charts 4 to 7). Overall, the story is of continued growth, perhaps with some deceleration. The one blot on this is unfortunately

Chart 4 OECD composite lead indicator for Indonesia

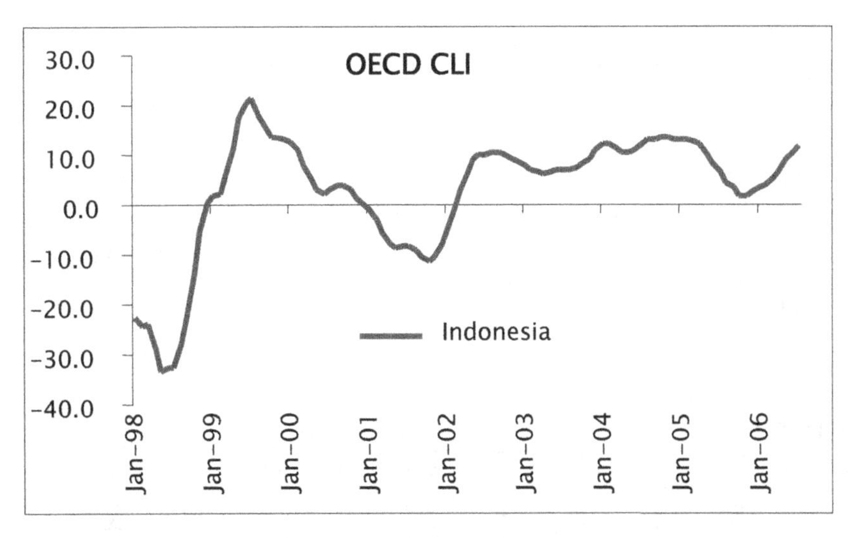

SOURCE Collated by Centennial Group using OECD Database.

Chart 5 Singapore composite leading index

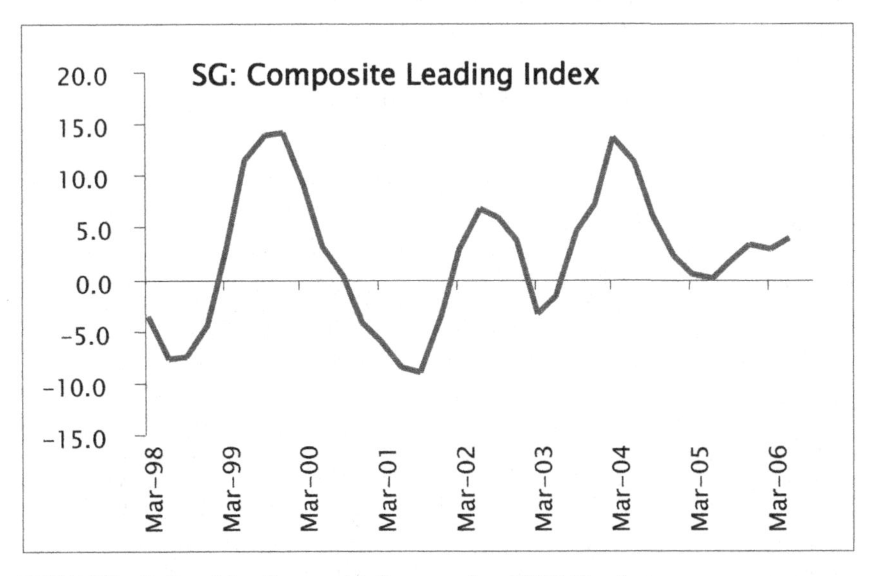

SOURCE Collated by Centennial Group using CEIC Database.

Chart 6 Malaysia's leading composite index

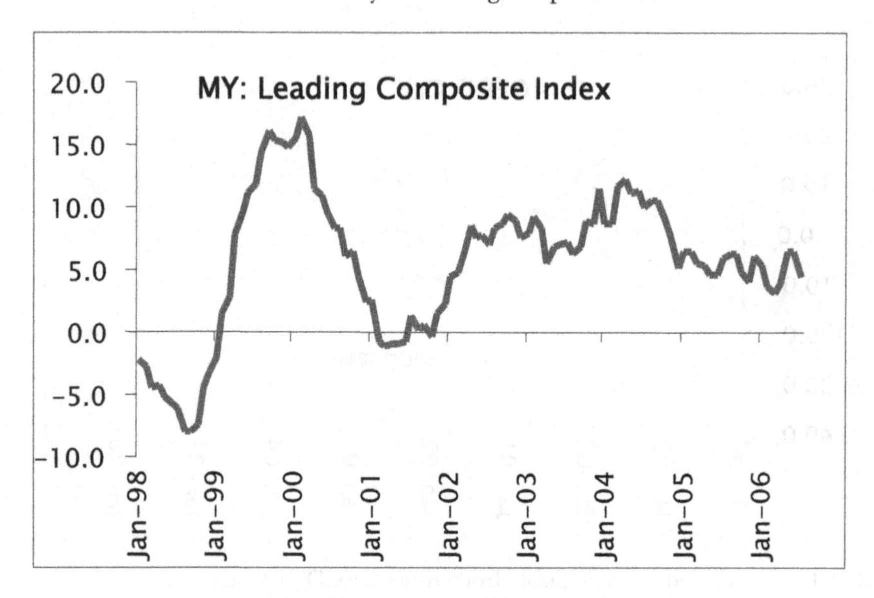

SOURCE Collated by Centennial Group using CEIC Database.

Chart 7 Bank of Thailand's leading index

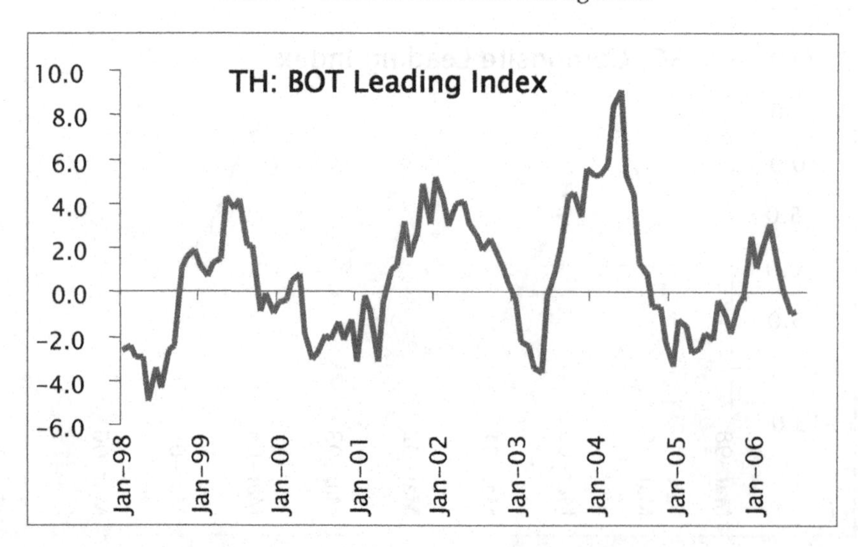

SOURCE Collated by Centennial Group using CEIC Database.

Thailand (Chart 7), where we have had a series of unpleasant surprises that make this picture probably a lot worse in the very near term of the first quarter. But I have always been more optimistic about Thailand than most people. Sometimes I am wrong, but I will stick my neck out to say that after surprising us unpleasantly in the last few months, Thailand over the next 12 months will surprise us positively.

Domestic/New Engines of Growth Emerging

The other positive is to be found in the domestic engines of growth: there are new, emerging domestic sectors as well as older sectors which are seeing a revival. An example of the latter is the construction sector, which is partly driven by the property sector. This sector has been in decline for a decade. Now, we appear to be over that phase, partly because the property sector is turning around and partly because the shake-out in the construction sector is over. This is an important trend because it will bring about a lot of multiplier effects to the economy. What is also encouraging is that there are a lot of new services, high-value added services that are emerging: IT services, creative industries, and exportable services in education and medical services.

Chart 8 Property turnaround

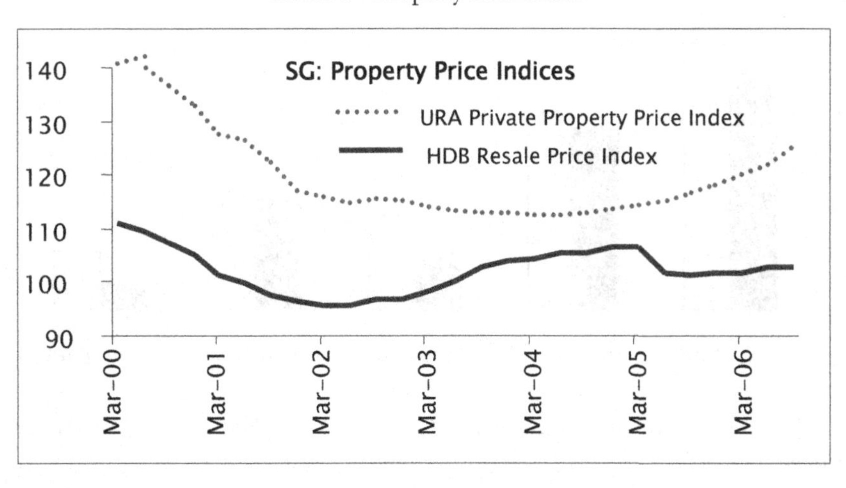

SOURCE Collated by Centennial Group using CEIC Database.

As can be seen from Chart 8, the property price indices are beginning to recover, particularly private residential properties. What is most important is that the stock of unsold housing is coming down significantly. That means that the property recovery will broaden out to help larger segments of the property market and therefore boost the wealth of a larger part of the consuming public in Singapore.

Singapore Now More Resilient

In the light of the unanticipated surprises in the global economy and in the region that I mentioned earlier, it is important to understand how resilient the Singapore economy is to these shocks. And the good news is that Singapore has actually become a lot more resilient over the last few years (see Chart 9). This conclusion is arrived at by combining various components of what drive economic resilience into an index and comparing it over time. Singapore has improved significantly in this measure of economic resilience.

Chart 9 Singapore tops in resilience

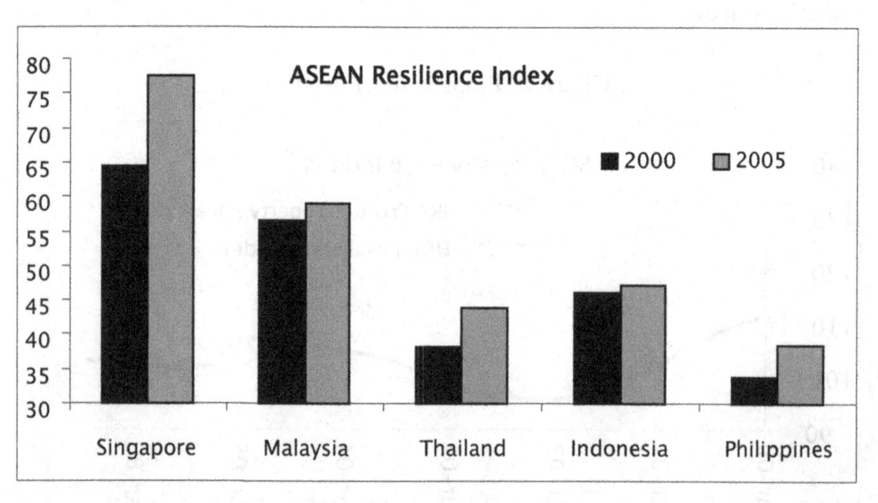

SOURCE Collated by Centennial Group. A higher reading indicates stronger resilience.

The reason for this is, first of all, greater diversity in the economy. A more diverse economy does not depend just on one engine of growth. If this is the case and that one engine sputters, then the economy will plunge. In Singapore's case, we now have more engines of growth kicking into life: construction, the domestic sector, and a manufacturing sector that no longer just depends on electronics but also on pharmaceuticals and transport engineering, thanks to the likes of Keppel and SembCorp. We also have a lot of momentum in the system and that helps offset any near-term economic weaknesses. Our financial sector is much stronger in terms of the balance sheets of the banks and of non-performing loans. The external position, driven by huge surpluses, gives us a lot of buffer (see Charts 10 and 11). Overall, the government's capacity to respond, of course, is strengthened by our physical reserves and the more flexible economy that we now have.

Chart 10 Singapore's trade balance

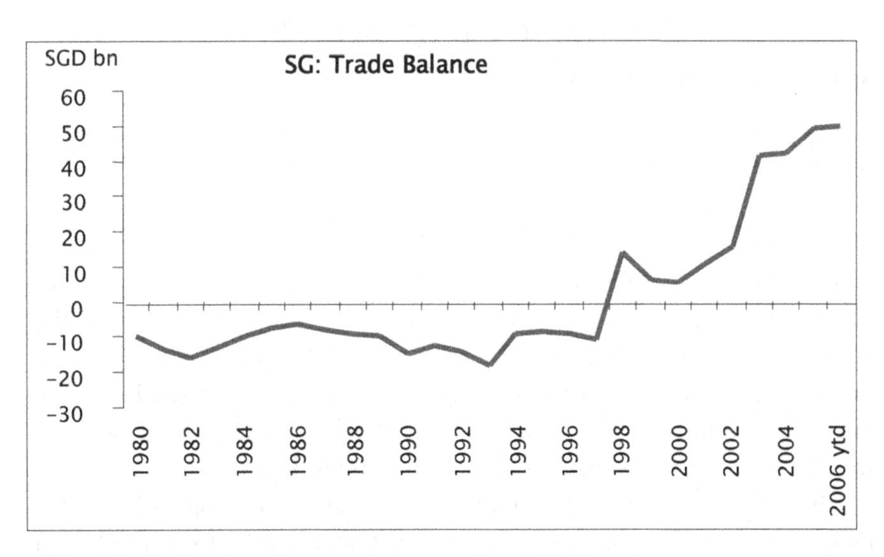

SOURCE Collated by Centennial Group using CEIC Database.

Chart 11 Rising trade balance

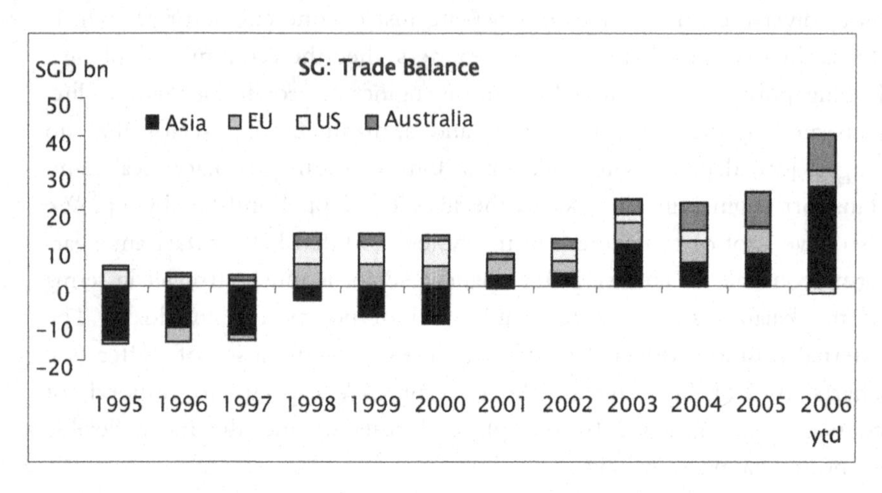

SOURCE Collated by Centennial Group using CEIC Database.

LONGER-TERM PROSPECTS

Strong Push for Growth

Singapore's longer-term prospects are boosted by the improving prospects for its regional hinterland. The ASEAN economies, after having gone through a very difficult phase for the almost ten years since the Asian financial crisis, is now ready for a new burst of growth, driven by a rising investment cycle. That is very important for us because it will spill over into demand for all kinds of goods and services that Singapore provides to the region. It will help boost the financial centre here because Singapore will probably be a major player in the capital raising needed to fund the new infrastructure and other projects. We are also extremely lucky to lie between the two large emerging giants of Asia: China and India.

In addition to the regional hinterland, we are also going to be driven higher by the inflow of talent. And we will also benefit from the lagged effects of deregulation, which has unleashed entrepreneurial energy in Singapore.

ASEAN in New Growth Phase

Let me make a case for ASEAN. The grouping has done a lot in the last ten years. It is not just a case of absorbing the shock from the series of financial, currency and political crises that visited us in the late 1990s. Southeast Asian economies had to deal with more than that. There really were two shocks in the late 1990s. First was the financial shock following the collapse of regional currencies in 1997. But the second shock was that, at the same time, China's emergence was unleashing a huge trade and competitive shock as well. ASEAN has now weathered both these shocks. The member countries have readjusted, re-engineered and they are prepared for a new phase of growth.

Second, the long period of under-investment in the region, which saw the ratio of investment to GDP across ASEAN falling sharply, is now probably also over. Rising investment is likely to be driven, not just by major infrastructure projects but also by the return of foreign direct investment (FDI). For instance, data just released show that Thailand actually was the biggest winner across the world in terms of the pace of increase in FDI in 2006. In 2006, there was a 114 per cent rise in FDI flows into Thailand.

Third, we are seeing the development of new growth poles. The Iskandar Development Region in south Johor, which is actually a very well thought-out strategy, can immeasurably help boost Singapore's prospects — if we rise to the opportunities presented by this new region. In addition, there are also the new special economic zones in Batam and Bintan which are likely to revitalise those neighbouring regions. These developments — in south Johor as well as in our nearby island neighbours — put us at the centre of two big potentially dynamic and vibrant regions which will clearly give Singapore a lot of scope to export our goods and services. But more than that, they could potentially help Singapore expand its economic space beyond our political boundaries, in turn releasing huge synergies with benefit to Singapore and its neighbours. These are tremendous positives.

Charts 12 and 13 substantiate some of my points. As you can see from Chart 12, in terms of shares of world export, ASEAN has continued to maintain a very high share of global exports even as China increased its share. In other words, China's increased competitiveness did not come at the expense of ASEAN. That is very important.

Chart 12 Southeast Asian competitiveness

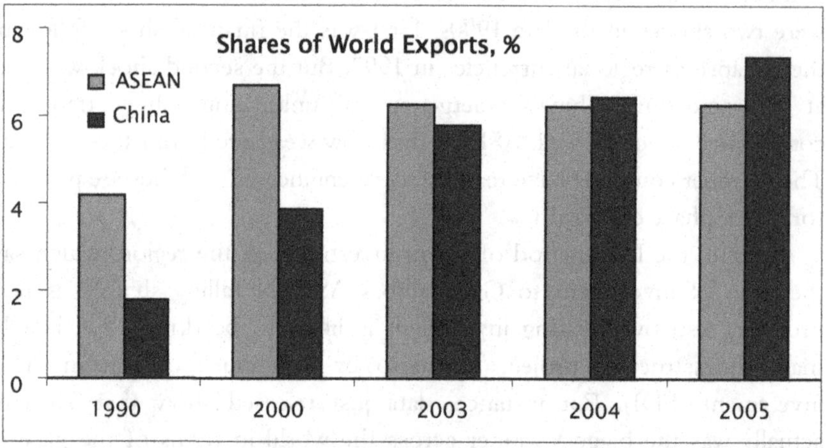

Shares of World Exports, %
ASEAN
China
8
6
4
2
0
1990
2000
2003
2004
2005

SOURCE Collated by Centennial Group using WTO Database.

Chart 13 Share of FDI to developing countries

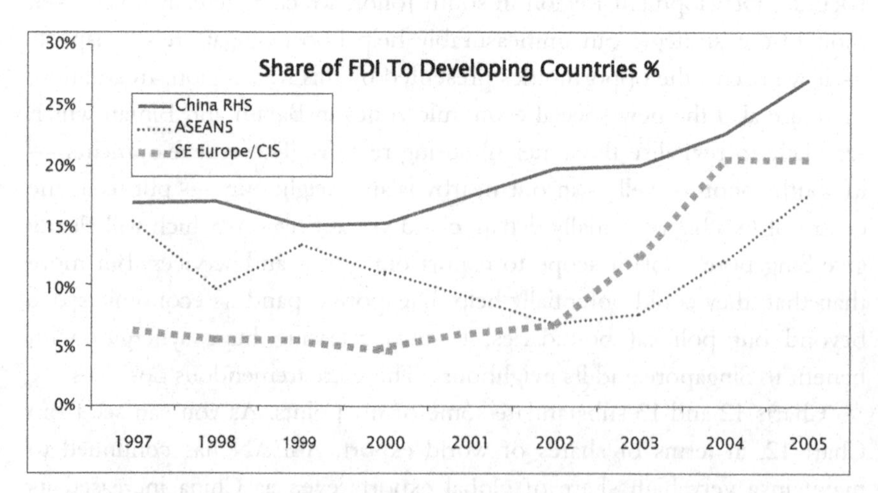

SOURCE Collated by Centennial Group using UNCTAD Database.

Chart 13 looks at the share of FDI flowing into developing economies. Here, too, it is not just a question of China but a question of all kinds of countries now competing for FDI. Look at how emerging Europe is as much a competitor for FDI as China is. Again, if you look at the five core ASEAN economies, they have, after a very difficult period, started to receive a fairly good share of global FDI.

Consider also new areas of growth. In the last couple of surveys done by A. T. Kearney on global services competitiveness, ASEAN economies have zoomed up, some from nowhere (see Table 1). For instance, Indonesia has gone from being unranked to the top ranks of countries whose economies can provide outsourcing, offshoring and the like.

Table 1 ASEAN and services growth

	2005	2004
India	1	1
China	2	2
Malaysia	3	3
Philippines	4	6
Singapore	5	5
Thailand	6	13
Czech Republic	7	4
Chile	8	9
Canada	9	8
Brazil	10	7
Indonesia	13	Unranked

SOURCE Collated by Centennial Group using A. T. Kearney reports (various years).

And of course there is the impact of China and India. Chart 14 shows that Singapore's trade balance with both countries is improving. In other words, they are adding very much more to our economic growth.

Chart 14 Impact of China and India

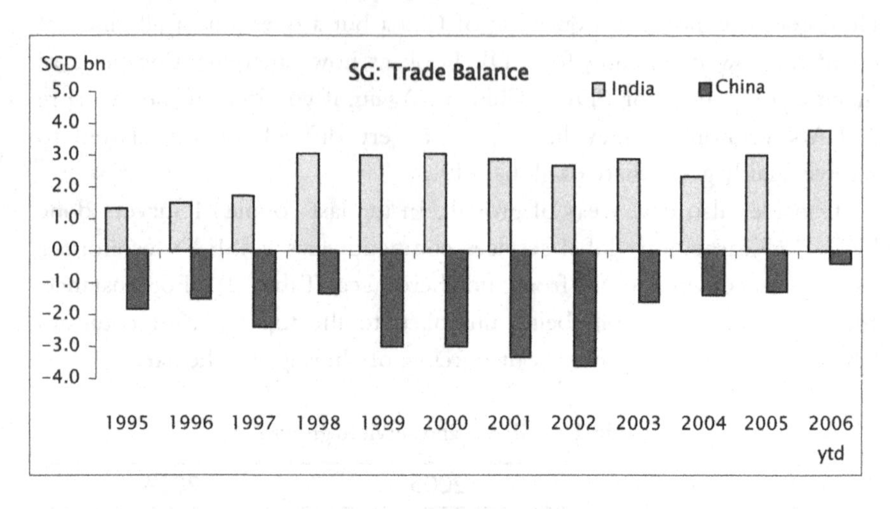

SOURCE Collated by Centennial Group using CEIC Database.

Singapore's Talent Capital

Singapore's aggressive drive to attract foreign talent rests on the call to develop a "talent capital" to steer its economy towards a highly specialised, innovative and knowledge-intensive one characterised by cutting-edge research, niche marketing and techno-capitalism. Numerous benefits can be reaped from the pursuit of this policy stance. First and foremost, the inflow of foreign talent (see Chart 15) can ease the negative impact arising from the country's limited pool of local talent and a crippling labour force (a result of declining fertility rate over the years). In addition, this pool of foreign talent can bring in high niche soft skills and innovative ideas to facilitate knowledge creation in the economy and the evolution of a pool of high-quality local talents and leaders to steer the economy to be innovative, competitive and entrepreneurial globally.

Meanwhile, established MNCs globally will also be attracted to the country's pool of high-quality labour force and will tend to locate their high-value work in Singapore, hence boosting the domestic economy. Not forgetting the potential spending power that foreign expatriates can contribute to domestic consumption, their demand for high-end residential housing will also give the construction industry a boost. Another advantage includes the possible strengthening of the country's fiscal position attributed to the robust inflows of tax revenues from these growing numbers of foreign expatriates and MNCs being drawn into and located in the country. In sum, this pool of talented specialists, also known as "shape-shifting portfolio people", is perceived to be crucial in transforming Singapore's small and open economy (with limited resources) to a high-end knowledge and service hub that continues to progress amid growing global competition.

Chart 15 Growing pool of foreign talent

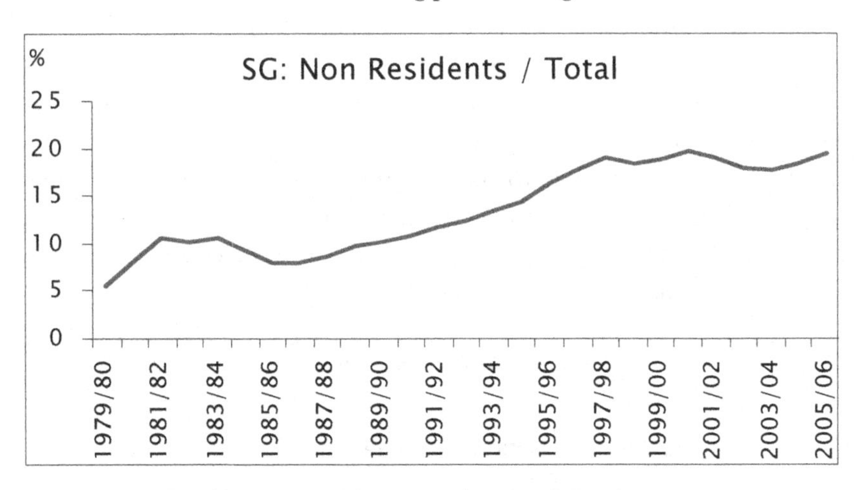

SOURCE Collated by Centennial Group using CEIC Database.

New Engines of Growth in Singapore Economy

In terms of engines of growth, we are not just an airplane with one or two small engines. We now have five or six engines powering us, so even if one drops off, we are still going to do quite well. We have pharmaceuticals and high-end electronics in manufacturing. We strongly disagree with the talk about how Singapore must somehow become like Hong Kong and give up manufacturing because, it is averred, we can never be competitive in manufacturing. We have a very strong position in high-end manufacturing and I think it is good for us to have a diverse range of growth engines, unlike Hong Kong. Also in services, we are seeing a boom in wealth management, in oil-related services and in IT services very distinct from the hardware manufacturing components of the industrial sector. There are also encouraging signs in the growth of creative industries.

Areas of Concern: Who is this Growth for?

The prospects therefore sound extremely good. But, I think I need to balance that with some important questions. Who is this growth for? The widening inequality in terms of measures such as the Gini coefficient or the ratio of wages over GDP (see Chart 16) is worrying. Chart 16 shows the diverging trends in who gets a bigger share of the cake — labour or capital — and it is clear that the corporate sector rather than labour has been the main winner in this phase of growth. It suggests that there are many losers in this new phase of growth, and not everyone has benefited from the growth. The problem is not that one group of people is seeing slower growth of income compared to another group. What is worrying is there is a group of people in Singapore, and quite a large one, whose real income has actually fallen in the last few years. And this is much more worrying than just saying that the gap has widened.

Chart 16 Wage share falling

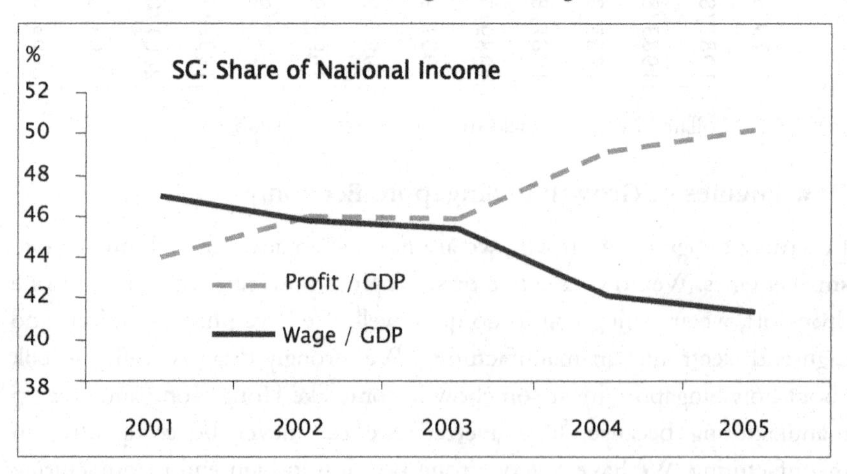

SOURCE Collated by Centennial Group using DOS Database.

More Losers in New Phase of Growth

The Gini coefficient has increased significantly just in the last few years (Chart 17). And as you can see from Chart 18, the average household income in the key 11th to 20th percentile has actually fallen. This is very worrying. These are nominal figures, and after adjusting for inflation, the picture is much worse.

Chart 17 Increasing Gini coefficient

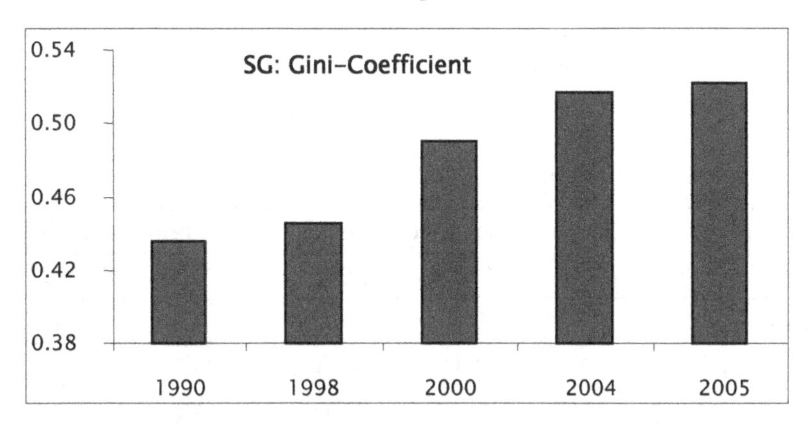

SOURCE Collated by Centennial Group using DOS Database.

Chart 18 Declining average household income

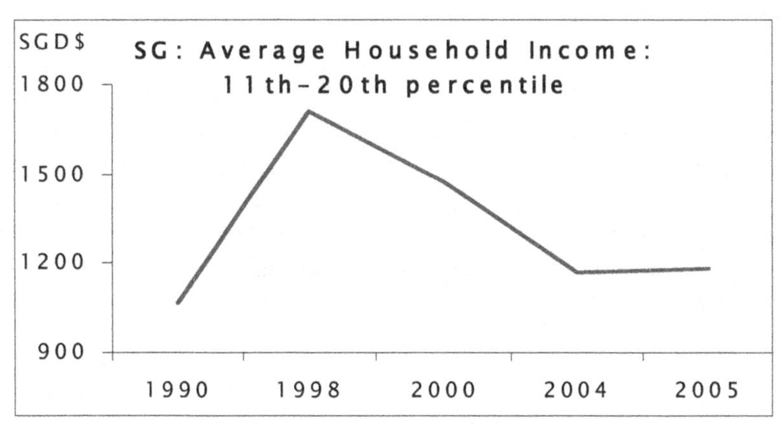

SOURCE Collated by Centennial Group using DOS Database.

Local Business Returns

It is also important to look at the performance of local companies as opposed to foreign companies. Local businesses too tend to underperform in a whole range of corporate measures. Figures on the returns on capital of local businesses versus foreign businesses show that local businesses have consistently underperformed (Table 2). That pattern has continued and it is a pattern that is found across every sector. So again who benefits from the growth and what is it in the structure of this economy that somehow militates against indigenously-owned businesses doing well? This is not a thing that we can sweep under the carpet because it has very important social, political and economic implications.

Table 2 Local business returns

	Local-controlled		Foreign-controlled	
	1991	2003	1991	2003
Total	**9.4**	**9.3**	**24.4**	**14.5**
Financial services	7.1	11.4	26.7	10.1
Non-financial services	11.4	7.1	23.4	17.1
Manufacturing	12.5	6.8	25.1	20
Construction	9	–9.2	8.5	–4
Commerce	12.2	5.1	33.4	15.9
Transport & communication	16.3	11.9	8.1	20.4
Insurance services	5.2	23.5	17.1	33.9
Real estate & business services	8	3.2	4.8	3.9

SOURCE Collated by Centennial Group using Singapore Corporate Sector Report (various years).

CONCLUSION

Singapore's growth prospects are good. The regional and domestic engines are reviving. Even if some short-term risks materialise, the Singapore economy is likely to be remarkably resilient. Only if the US economy does really badly and suffers a serious recession will Singapore suffer a downturn as severe as, say, 2001.

The policy challenge is therefore not necessarily to maximise economic growth at any cost. The really important policy question is: "What is the aim of all this growth?" Is it just to maximise GDP growth? Or is it to ensure that as large a proportion of Singapore citizens as possible benefits from economic growth? In other words, it is the quality of that growth which might be more important than the level of growth.

SECTION **2**

The New Politics of Identity

CHAPTER 4

Migration and Social Diversity in Singapore[1]

BRENDA S. A. YEOH

INTRODUCTION

Singapore as a small, natural resource-scarce city-state has been part of global enterprise and global service capitalism since its very birth. As a child of imperial and labour diasporas in the colonial era, our story is very much intertwined with migration and migrants. An openness to migrants is how we have gained importance and stature in the world. With the colonial era behind us, we are a "wannabe" global city, competing very hard for a place in the top league of globalising nations. Once again we are a convergence node for transnational flows of different sorts of migrants from many parts of the world.

A migrant history is thus where we come from and that gives us, hopefully, a sense of empathy for migrants. We know the inside story of migration because it is part and parcel of our being, our histories, our families perhaps, and our place in Singapore today. And as part of the story of migration, a dependence on foreign labour is also very much part and parcel of our economic survival through the times, from the colonial era right up to today.

Our economy today is restructuring and transiting to services, financial and high-technology areas. Migrant labour continues to be crucial. The Ministry of Manpower puts it in a nutshell when it says in the Manpower 21 statement that the employment of foreign manpower is a "deliberate strategy to enable us to grow beyond what our indigenous resources can produce". In Singapore, a city-state that is very well managed in many regards, leveraging on migrant labour is part of this carefully managed, "deliberate strategy" of growth.

CHANGING POPULATION COMPOSITION

So what implications do the increased flows of migrants have on the social landscape of Singapore? Let us first take a look at Table 1 which shows population figures compiled from census reports.

Table 1 Changing proportion of citizens to non-citizens in Singapore

Census Year	Citizens		Permanent Residents		Non-resident Population		Total	
	No.	%	No.	%	No.	%	No.	%
1970	1,874,778	90.4	138,785	6.7	60,944	2.9	2,074,507	100
1980	2,194,280	90.9	87,845	3.6	131,820	5.5	2,413,945	100
1990	2,595,243	86.0	109,872	3.6	311,264	10.3	3,016,379	100
2000	2,973,091	74.0	290,118	7.2	754,524	18.8	4,017,733	100

SOURCE Adapted from Yeoh, 2006, page 28, Table 1.

As of the year 2000, the non-resident population makes up 18.8 per cent of a population of just over four million, in other words, one in five people on this island. If we focus on the division between citizens and non-citizens instead, then basically every one in four people on the streets that we meet today is a non-Singaporean.

Turning now to changes over the decades, it is obvious that the non-resident population is the part of the population that is growing in the most rapid fashion, almost doubling every decade in terms of its percentage of the total population. In short, from the 1970 census to the 2000 census, the population of Singapore has doubled from about two million to four million. The non-resident population has grown from 2.9 per cent in 1970 to a hefty 18.8 per cent of the population of four million today. The non-citizen population has likewise grown from 9.6 per cent in 1970 to 26 per cent in 2000. The 2005 figures also show a further increase of the non-citizen population to 28.4 per cent out of a total population of 4.3 million.

Table 2 Resident population by ethnic group and residential status

	Total	Chinese		Malays		Indians		Others	
Total	**3,263,209**	**2,505,379**	**76.8%**	**453,633**	**13.9%**	**257,791**	**7.9%**	**46,406**	**1.4%**
Singapore Citizens (91.1%)	2,973,091	2,284,617	76.8%	441,737	14.9%	214,642	7.2%	32,095	1.1%
Permanent Residents (8.9%)	290,118	220,762	76.1%	11,896	4.1%	43,149	14.9%	14,311	4.9%

SOURCE Department of Statistics, 2000.

In terms of ethnic composition, the CMIO (Chinese, Malays, Indians and Others) model that is foundational to Singapore's multi-ethnic philosophy remains more or less in place despite the decades of in-migration. However, there are dips in the Chinese and Malay percentages and increases in the Indian and Others percentages. This may be accounted for if we look at the differences in the ethnic profile between Singapore citizens and Permanent Residents (see Table 2). Among the Permanent Residents, the Indian population is 14.9 per cent, which is twice the percentage of Indians within the Singapore citizen pool. The Others category, of course, features as a larger percentage of the Permanent Resident population than of the Singapore citizenry.

"NEW" SOURCES OF SOCIAL DIVERSITY

Clearly, then, we have been an immigrant society of considerable diversity since our beginnings as a port-city. Today, under conditions of globalisation, what sorts of new social diversity do we confront? As Singapore fashions itself as a magnet for foreign manpower, we have in our midst different categories of migrant labour. "Foreign workers" is the term we tend to use to refer to the large numbers of low-skilled labourers in construction, manual work, cleaning and domestic work. A category that is much smaller in number but of importance is skilled labour — employees at the professional and managerial level, whom we commonly refer to as "foreign talent". A third group that is increasingly important — because they are the future manpower — and growing in numbers is the foreign students. At all levels there has been an increase in foreign students in our schools and institutions of higher learning. In sum, the non-resident workforce of over 600,000 in 2000 makes up about 30 per cent of total employment in Singapore, the highest figure in Asia.

Managing "Foreign Workers"

We first turn to the category of foreign labour that has the largest number. Foreign workers, who number more than half a million in the industries mentioned before, come from a wide range of countries and include Chinese, Indians, Filipinos, Thais, Malaysians, Indonesians, and Sri Lankans. These workers perform the kinds of work that Singaporeans themselves have been reluctant to do.

Our changing social landscape shows signs of their presence. We see them in Little India, the Golden Mile Complex, Orchard Road, and other public places when they leave the confines of construction sites, workers' dormitories and employers' homes on their days off.

On 3 March 2002, a photograph which appeared in *The Straits Times* showed a scene rarely seen in Singapore. A group of Chinese nationals working in the construction industry had staged a protest outside Parliament House to remonstrate against what they felt was the slow pace of investigations into the case of a remittance agent who fleeced them of large sums of money. Riot police had to be deployed to break up the protest — certainly something of a head-turner in Singapore, for how many Singaporeans would stage a protest outside Parliament House to demand justice? This is perhaps a sign that things are changing, not just in the economic sphere, but also in the socio-political sphere.

Socio-political change is, however, also kept carefully in check. Foreign workers are "managed" through a whole series of control measures — including the Work Permit system, the dependency ceiling, government levies and the security bond — to ensure that they remain a transient population, carefully regulated in order to avoid what may seem to be disruptive social effects. State policy is opposed to the long-term immigration of low-skilled workers and is directed at ensuring that they remain temporary and easily repatriated in times of recession. As part of this overall policy of transience, family formation is circumscribed, dependents are barred, and marriage to Singaporeans or Singapore Permanent Residents is not allowed. And for the women — that is, foreign domestic workers — getting pregnant amounts to repatriation. In short, state policy treats foreign workers as disposable labour which must not remain threaded into the basic fabric of Singapore society.

Attracting "Foreign Talent"

Cities which are seriously competing in the globalising game are very much concerned with what has been called a "global war for talent". Singapore is no exception and there is an all-out effort to build Singapore into a "brains service node", an "oasis of talent", or the "Talent Capital of the New Economy", to use some of the catchphrases contained in government documents and speeches. While foreign talent comes from a wide spectrum of countries, numbers from China and India in particular have been growing rapidly in the last decade or so.

Major initiatives have been launched to attract, retain and absorb foreign talent, such as the liberalisation of immigration policies, recruitment missions, Permanent Residency schemes and company grant schemes which ease the cost of employing skilled labour. The whole re-imaging of the architectural and cultural frameworks of urban development is yet another major initiative that serves the purpose of attracting and retaining talent. Efforts to remake Singapore into a "Global City for the Arts" are at least partly aimed at retaining foreign talent who might not want to stay on if Singapore remains a cultural desert.

Skilled individuals in arenas ranging from science and technology to sports — including their family members and dependents — are hence part and parcel of our everyday social landscape. Living in a globalising city means that we encounter these individuals regularly beyond the workplace, on the streets, in shopping centres, schools and so on.

Investing in "Foreign Students"

Another source of social diversity takes the form of international students as Singapore seeks to fashion itself as an education hub for the region. The "Global Schoolhouse" project, as it is called, involves creating Singapore as a world class educational hub, known for its intellectual capital and creative energies. Investing in foreign students is another way of augmenting our pool of skilled manpower for the near future. This is a strategy that affects the whole education landscape from schools to universities. Here, Singapore is capitalising on its various strengths, including its English-speaking environment, high standards of education, low crime rates, high order of social discipline, as well as what has been called our "squeaky clean", "nanny state" reputation, to make parents feel assured about sending their children here.

The aim is to more than double the number of foreign students coming to our schools and universities from about 66,000 in 2005 to 150,000 by 2012. The main targets are China and India as well as neighbouring Southeast Asian countries.

Turning to "Foreign Spouses"

Social diversity in Singapore is not just a product of the nation's labour policy. While manpower needs are the main drivers for the increasing social diversity we are experiencing in the population, there are other pressures at work. One of these has to do with the falling marriage rates in Singapore.

As seen in Table 3, the proportion of singles has gone up over the decade (1990–2000) in nearly all age groups among both Singapore citizens and residents. The increase in the rate of non-marriage as well as delayed marriage has been the subject of much public discussion as well as the source of state anxiety in recent years. Various reasons have been propounded, including the hectic pace of life in a globalising city that squeezes out time for love, sex and marriage, as well as the higher levels of education, financial independence and career-mindedness among Singaporean women. Social perceptions of singlehood are apparently also changing, becoming more acceptable than before.

Table 3 Proportion single by residential status, sex and age group

Per Cent

Age Group	Singapore Residents		Singapore Citizens	
	1990	2000	1990	2000
Males				
20–24	94.2	95.2	94.1	95.4
25–29	64.1	64.2	64.0	66.4
30–34	34.0	30.7	33.9	33.3
35–39	18.1	19.7	17.8	21.5
40–44	10.9	14.8	10.4	15.5
Females				
20–24	78.5	83.8	79.0	86.6
25–29	39.3	40.2	39.6	45.5
30–34	20.9	19.5	20.9	21.9
35–39	14.8	15.1	14.7	16.2
40–44	11.5	13.6	11.4	14.1

SOURCE Singapore Department of Statistics, 2000.

Linked to these trends, Singaporeans searching for suitable marriage partners are increasingly turning to sources from abroad. As with globalising cities elsewhere, for example, Taiwan, Japan and South Korea, turning to "foreign brides" from the region appears to be a growing trend in Singapore. Foreign spouses are hence an important source of social diversity that affects the basic social fabric of Singapore.

According to the Department of Statistics, there were 8,116 marriages (35.3 per cent) between residents and non-residents in 2005. This can be broken down into 6,520 male Singaporeans and Permanent Residents wedded to foreign brides and 1,596 of their female counterparts marrying foreign grooms. As the actual nationalities involved in these cross-nationality marriages are not released to the public, one can only speculate that the foreign brides are largely Chinese, Indian or Southeast Asian, perhaps, Vietnamese, Filipinos and so forth. Looking at the websites of matchmaking agencies, it appears that marriages to foreigners can be facilitated by packaged tours to Vietnam, for instance, to handpick one's partner from a wide range of applicants.

CURRENT DEBATES – MIGRATION AND INCREASING SOCIAL DIVERSITY IN GLOBALISING SINGAPORE

Migration, and the increasing social diversity it brings, have been very much in the news in recent years. Prime Minister Lee Hsien Loong gave these issues a central place in his 2006 National Day Rally, emphasising a three-pronged approach: encouraging procreation among citizens, engaging overseas Singaporeans in order to help them fit in when they return, and attracting foreigners who can contribute to Singapore to work, live and settle in the city-state. Apparently, efforts to recruit and retain foreigners and convert them to Singaporeans are bearing fruit because in 2005, there were 12,900 new citizens, compared to the usual 6,000 to 7,000 new citizens in the last four years before 2005. This gain seems substantial compared to the average figure of 800 Singaporeans giving up their citizenship each year.

SINGAPORE AS PERMANENT MAGNET FOR TALENT OR TRANSNATIONAL REVOLVING DOOR?

A key issue I would like to raise for discussion relates to the question of the permanence of the foreign population within our workforce. There is, of course, a major differentiation between the unskilled workers and elite transnational workers. For the unskilled workers, the revolving door is one that turns systematically and continually. "Use and discard" is the underlying philosophy behind a range of policies such as the Work Permit system which have been put in place to prevent unskilled workers from gaining a foothold in Singapore as socio-political subjects. Basically, we extract their labour for a certain number of years and then we send them back to their home countries.

In contrast, among the elites with talent and skills, the state tries to slow down this revolving door as much as possible and in fact encourages them to put down roots in Singapore, if not by becoming citizens, then as Permanent Residents, or at least as individuals willing to work in and contribute to Singapore for a good number of years.

Studies, however, seem to indicate that international talent is highly mobile. Even when Permanent Residency status and Singapore citizenship are used as carrots, they will not necessarily guarantee the integration of talented individuals into the fabric of Singapore society. In fact, attaining Singapore citizenship or Permanent Residency may confer a higher degree of potential mobility on them, enabling them to gain entry more easily as tourists and immigrants in other gateways around the world. In fact, people in a globalised world live transnational lives, negotiating a whole range of gateways and taking on multiple identities as part of the strategy to cope with the conditions of globalisation. Globalising cities such as Singapore are thus sites with a very high density of interactions between all kinds of locals and transnationals, not just in the economic sphere, but in all spheres at all scales, from the family to the nation-state.

In this globalising city, some key questions pertain to how best we cope with living in the kind of society where constant mobility is taken for granted. In this context, are we are fashioning Singapore as a transnational revolving door or an immigrant gateway? And if we want Singapore to be a gateway where people come and stay and contribute, is this to be done on a highly selective basis, and if so, what will the selection criteria be based upon? Is the globalising city only a place of welcome for those with talent and skills? How can we manage this divide mentality between welcoming the skilled and guarding against the unskilled and, at the same time, pay heed to upholding high standards of human rights and dignity congruent with a developed nation like ours? How do we ensure that the way we treat foreigners in our midst is not just calibrated according to, and polarised by, the market value of their skills? We need to remember that when we import a human being to work in our country, we are not just importing a skill or labour power, but socio-political subjects who carry with them their own cultures, values, hopes and dreams.

There is thus a need to take a harder look at the social integration and support mechanisms that we have, not just for the talented but also for all levels of skills, and perhaps evolve an integration policy which matches the complexity of our immigration and labour policies. While we have carefully calibrated immigration-labour categories, we also need an equally complex way to integrate human beings of all skill levels after they have entered Singapore. Integration and support mechanisms should go beyond foreign talent to include foreign students, foreign workers and foreign spouses, and each category deserves careful consideration for better-tailored support mechanisms. For example, there are hardly any support groups for foreign spouses who arrive in Singapore and who do not speak any of the mainstream languages here.

Finally, it leaves me to stress that debates about migration and migrants are significant, not just because of their importance in the labour force, but certainly also because of the way migrants of all categories shape the contemporary social landscape, as well as our demographic future.

ENDNOTES

1. Parts of the discussion in this paper can be found in expanded form in Lam, T., Yeoh, B. S. A. and Huang, S. " 'Global Householding' in a City-State: Emerging Trends from Singapore," *International Development Planning Review*, Vol. 28, No. 4 (2006) and Yeoh, B. S. A. "Singapore: Hungry for Foreign Workers at all Skill Levels," *Migration Information Source*, January 2007. Available online at http://www.migrationinformation.org/Profiles/display.cfm?ID=570.

REFERENCES

Singapore Department of Statistics. "Singapore Census of Population, 2000. Advance Data Release No. 8: Marriage and Fertility," accessed 28 February 2007 from http://www.singstat.gov.sg/papers/c2000/adr-marriage.pdf.

Yeoh, B. S. A. "Bifurcated Labour: The Unequal Incorporation of Transmigrants in Singapore," *Tijdschrift voor Economische en Sociale Geografie (Journal of Economic and Social Geography)*, Vol. 97, No. 1 (2006): 26–37.

CHAPTER 5

Just What is *New* about the "New Politics of Identity"?

KWOK KIAN WOON

IDENTITY IS ALMOST ALWAYS POLITICAL

The theme that we have before us is: "The new politics of identity". In what follows, I attempt to address the question: just what is *new* about the so-called "new politics of identity"?

To begin with, I propose that whenever "identity" is raised as an issue, whether by individuals, groups or institutions, it is almost always *political.* It is political in the general sense that an explicit concern with "identity" tends to be motivated by specific *interests*, especially by a combination of what the classical sociologist Max Weber called, on the one hand, "ideal interests" (e.g. keeping true to certain religious or political ideas) and, on the other hand, "material interests" (e.g. securing economic advantages or maintaining class positions).

Put simply, identity — especially Identity with the capital "I" or, as I sometimes like to say, "Eye") — has to be seen, projected, asserted, displayed, demonstrated, and defended in relation to "Others", most clearly when this is done in an "us *versus* them" manner. And such processes of purposeful *identification* tend to be manifested at a time when identity, which one lives out naturally — that is, one's sense of self or group identity interwoven seamlessly into the fabric of everyday life — can *no longer* be taken for granted. For example, individuals and groups may find themselves placed in a new environment which destabilises their positions in society or makes them feel oppressed in one way or another. In other words, "the politics of identity" can be framed as the politics of *recognition*, which also implies non-recognition or misrecognition. Issues of identity and identity politics emerge when specific persons or groups experience, perceive and articulate a sense of being *not* recognised or being *mis*recognised, that is, recognised by others in ways which are negative. Indeed, the politics of identity invariably involves the use of stereotypes among groups, each drawing invidious contrasts and thereby marking clear boundaries vis-à-vis others.

The questions that follow, therefore, are: what are the stakes in the politics of identity, and for whom? Often, identity or identification becomes more consciously thought out and fought over by *minorities*. This is rooted in the asymmetrical relationship between the "majority" and the "minority". In some ways, members of a majority can take for granted their collective identity or personal identities in the way that their minority counterparts cannot. Sociologically, a "minority" is not necessarily defined by numbers: any group of any size can be "minoritised" (that is, marginalised) by others through discrimination or by itself through its own self-perception of being discriminated against. In responding to non-recognition or misrecognition, members of such a group, whether or not numerically dominant, may have specific interests in developing a heightened sense of identification and accentuating specific identity markers.

THE OLD POLITICS OF IDENTITY: THE PLURAL SOCIETY AND ITS LEGACY

In order to delineate what may be new about the new politics of identity, I shall attempt to address the question: what has been the old politics of identity? In a nutshell, the old politics of identity in Singapore has been framed by the legacy of the "plural society" of the colonial era. The concept was introduced by the British administrator J. S. Furnivall, who referred to societies consisting of "two or more elements of social order that live side by side and yet without mingling in one political unit". Furnivall was concerned about whether or not membership in the larger society — and not just within one community — created common social experiences and a "common social will" beyond production and market exchange. In his view, the heterogeneity of a plural society inhibits the development of a sense of "common citizenship"; social life is "sectionalised" and the potential for conflict among various sections makes the society inherently unstable — "enhancing the need for it to be held together by some force exerted from outside" (Furnivall, 1944; quoted in Hefner, 2001: 4–8).

In the course of decolonisation and the making of the independent nation-state, the political leaders in Singapore made a commitment to "multiracialism", which was, arguably, an extension of the colonial plural society. The earlier multiple categories of people from different races used in colonial administration became more managed under the broad categories of CMIO — Chinese, Malay, Indian and Others. These are the so-called four "races", although "Others" appears to be a residual category here. In lieu of a critical discussion of the People's Action Party (PAP) government's policy of multiracialism, which has been analysed by other scholars (Benjamin, 1976; PuruShotam, 1998; Chua, 1998; Tan, 2004), it may suffice to say that the CMIO classification as a model of managing ethnic relations — and identity politics — has tended to compartmentalise the four categories and to assume that there is an overlap between race, language and culture, solidifying the shifting and porous boundaries between groups. This is seen in a host of policies, including language and housing policies, and in the establishment of self-help organisations along racial or ethnic lines. The politics of identity, therefore, revolves, around claims for recognition and equal treatment among races.

While multiracialism has served as the predominant way of organising Singapore society, "multiculturalism" has emerged over the decades as an expanded concept, which gives way to the recognition of cultural and social attributes that go beyond race per se, for example, ethnicity, gender, age, class, sexuality, and religion. This is not unrelated to the widespread use of the term as part of official policy-making or public discourse in countries such as Australia and Canada. It also reflects the growing differentiation of the Singapore population into groups that cannot be simply defined by broad racial categories. Not surprisingly, the politics of identity along the lines of multiculturalism has tended to be "bottom-up", with claims of recognition made by non-governmental organisations. Hence, for example, the greater prominence of the women's and gay movements in asserting the principle of equal treatment accorded to the various races.

Multiculturalism is also accompanied by another kind of identity under the banner of "cosmopolitanism". With the acceleration of globalisation and the growing competition among cities as significant nodes in the flow of goods, services and talent, the government is keen to project the self-image of Singapore as a "global city" — a modern city-state that is open to international economic and cultural exchange. This image is rather glamorously projected in tourism advertisements and is part of the policy to retain and attract global talent. On the ground, too, there are sectors of the resident population that are more "globalised" than others by virtue of education, age, income and occupation and hence class; and they are potentially more mobile in the global marketplace. Not surprisingly, the new politics of identity revolves around issues concerning the relative benefits and freedoms enjoyed by locals and foreigners, by citizens and non-citizens.

TOWARDS A NEW CONFIGURATION?

Is Singapore moving towards a new configuration in the politics of identity? I would first pose the question whether, as much as we have moved beyond the plural society, there may still exist elements of the type of colonial-era plural society analysed by Furnivall. If we look at the different categories of residents and non-residents in Singapore in our midst — citizens, Permanent Residents, Employment Pass holders ("foreign talent"), Work Permit holders (overseas contract workers) — and the different kinds of mechanisms and logics of management for different groups, are there still elements of the old plural society at work? Taken together, these categories make up a sophisticated system of differentiation. As the credit card advertisement tagline goes, "membership has its privileges": there is differentiation between full membership, partial membership, and non-membership, the last of these being applied to transient workers, whose numbers can be adjusted according to the vicissitudes of the economy.

Singapore, however, has moved beyond the strict model of the plural society because social diversity has become more complex and more difficult to manage. The neat compartmentalisation of the plural society and multiracialism models has also given way to official efforts in partially dissolving the boundaries between groups, especially in the promotion of interracial, interreligious, and intercultural dialogue in the aftermath of September 11. In that sense, there is no way to return to the plural or multiracial society in which groups are like ships passing through the night, and the minimal condition is that they do not collide. And if there is likelihood that they do collide, the state steps in to use instruments of state coercion to control the situation.

"Race" is very deeply inscribed and institutionalised as an operating principle of social life in Singapore; even the policies for attracting Permanent Residents and hence also potential new citizens are guided in part by racial considerations. But racial identity for citizens and non-citizens is also giving way to many other kinds of identity. In the social science literature, we have concepts such as multiple identities, hyphenated identities, hybrid identities, and even situational identities. In any particular context, one can pick and choose, and mix and match different kinds of identities in this huge, global supermarket of identities to construct oneself, not least because of the overloaded floating world (or, indeed, worlds) of signs and images in the age of the Internet. As I have put it elsewhere: "I click, therefore I am." Another version reads, "I con, therefore I am" — playing on the icons found on the computer screen and the possibility of disguising and inventing virtual identities). And it is among the young that we are witnessing this kind of freedom, this kind of liberation, in the construction of self-identity. To put it another way, the "self" is no longer embedded in fixed categories, as fixed as, for example, in racial categories. The "self" is disembedded and set free — and this is embraced as a liberating experience by the young, but is perceived as a problem by those who wish to manage the new politics of identity.

IDENTITIES ARE NOW DE-TERRITORIALISED

What about the project of nation-building? Singapore has been called by Prime Minister Lee Hsien Loong in his 2006 National Day Rally speech an "improbable nation". Others have used terms such as "reluctant nation", "artificial nation", and even "freak nation". Was Singapore ever meant to be a nation? Does it really have to be a nation? Couldn't it just be one gigantic multinational corporation, with the Prime Minister as the Chief Executive Officer, the Cabinet as the Board of Directors, and citizens as the employees? Does Singapore really require Singaporeans? Such questions have been posed by scholars (cf. Kwok and Mariam, 1998). In the course of nation-building, however, with successive generations of young citizens reciting the National Pledge every day, the "imagined community" of Singapore as a nation has become a perceived reality in the minds of many citizens.

More concretely, citizenship confers rights and duties. Some of the tensions between citizens and foreigners have to do with the balancing of rights and duties, and the costs of citizenship, for example, in relation to National Service, education, and housing and medical subsidies. If Permanent Residents have the same, or more or less similar, rights without the same duties as citizens, the new politics of identity comes to the fore. The more globalised sectors of the population are also more able to "transcend" the nation, more able to say to Singapore, "Look, I am here but I don't have to be here, I can be in and out, I can be in or out." Identities are now de-territorialised, and if this is the case for citizens, it is even more so for foreigners in Singapore. At the same time, the less globalised sectors — and these, as mentioned, correlate with variables related to social class — are less physically and psychologically mobile. The new politics of identity, therefore, brings into relief a complex multiplicity of claims made by various groups, ranging from citizens who articulate a sense of being treated worse than Permanent Residents and foreign talent to those who are concerned about how transient workers are treated in the country.

WHAT IS UNIQUE ABOUT SINGAPORE CULTURE?

Economic motivations transcend identity, loyalty and attachment. The logic of economics implies that economically rational men and women will find their paths to places that enhance their economic interests. If so, thinking through the new politics of identity requires us to think beyond economics. What is so uniquely Singaporean that makes it so attractive and compelling for people — citizens, residents, as well as people who have come here as Employment Pass or Work Permit holders — to say: "This is a place that I can give a good part of my life to"? For want of a better term, we could say that that something is "Singapore culture" (and I use this as a term that connotes something wider and deeper than what political leaders have called "heartware" — that is, as distinct from economic "hardware" or infrastructure). But what is Singapore culture? In September 2006, Minister Mentor Lee Kuan Yew was reported to have predicted that "a Singapore culture is unlikely to emerge, not even in the next few hundred years" (*The Straits Times*, 6 September 2006). Of course, in anthropological terms, there is a Singapore culture and it has been shaped not just by the legacies of ancestral traditions and historical circumstance (including colonialism), but also by the making of the nation-state, especially through government policies and contemporary forces of globalisation.

MM Lee was also quoted as saying: "In the old days, meaning before this digital age, you had the time and the isolation to develop on your own and create something distinctive... Now you have to synthesise all the time. And out of this synthesis make something which is relevant to yourself and your future." He also said that Singapore does not "have the confidence to create its own culture", adding that "the basis of our culture is what we inherited from our original countries, our original cultures". I suggest that instead of thinking that Singapore's "own" or "distinctive" culture will never emerge, it may be more productive to think of "Singapore culture" as always in the making — and always involving processes of synthesis, including the kind of synthesis that MM Lee might have been referring to. Put simply, all kinds of syntheses have been tried out in Singapore and constitute what may be called Singapore culture. This is a very under-studied area. What is it about the Singaporean synthesis that is at once exciting and challenging, problematic and superficial; in the whole mix of Singapore culture, is there anything that we can say is deeply compelling, attractive and demanding of us the kind of commitment to belong to this place, this society, this nation?

CULTURAL DISJUNCTURE AND CREATIVE CITIES

I think the stakes are very high in the whole question of Singapore culture. Looking at the literature on the historical forces that have shaped the most significant cities of the world, we may draw some precious lessons. In particular, we may turn to Peter Hall's magisterial *Cities in Civilization* (1998), especially his analysis of the city as "cultural crucible". In the first pace, creative cities require "continual renewal of the creative bloodstream". This means that there is no way for Singapore to be xenophobic, for Singaporeans to shun foreigners and to be negative about this. We have to embrace this and confront the dilemmas and contradictions that accompany our openness to the constant renewal of the creative bloodstream — and, importantly, we also need to pay special attention to such renewal from within and not just from outside Singapore. Creative cities tended to have talented people who were experiencing and responding to the "throes of transformation in social relationships, in values, and in views about the world... they all were in a state of uneasy and unstable tension between a set of conservative forces and values... and a set of radical values which were the exact opposite". Now, there is no magic formula, no standard "essential ingredient" that can be used by policy makers in "engineering" the creative city. Hall (1998: 285–286), however, wrote: "What appears crucial is that this clash, at least this disjuncture, is experienced and expressed by a group of creative people who feel themselves to some important degree *outsiders*: they both belong and do not belong, and they have an ambiguous relationship to the seats of power and authority."

Hall's research also showed that "in every case, it seems that an initial economic advantage is massively transformed into a much larger cultural one". Now it seems to me that something of this nature is lacking in Singapore. If creativity is generated from periods of great transformations in social relations and value systems, we have seen such periods of creativity in our own history. And if there's any period during which many fundamental assumptions have come into question, we have entered such a period in the early 21st century — except that a lot of the questioning and a lot of the answering comes from official sources. In Singapore, the tension between the old and the new, the conservative forces and a newer set of values, this tension, is almost always managed in a neat and tidy way. I can't quite put my finger on it, but I think it is important to bring this up and put it on the table. There is every potential for Singapore to build on its economic advantage and to significantly transform it into a much larger cultural advantage. I believe that our policy makers aim to do this and are attempting to do this, but can we do it better, and how? Indeed, there may be no simple shortcut in engineering outcomes in this direction.

The basic point I wish to make here is that we need to understand how "cultural disjuncture" can be harnessed into creative energies, and we need to recognise that such energies can be generated from within and outside Singapore — thus breaking down of the invidious bifurcation between "foreign" and "local" talent. Allow me to provide two "local" examples of creative synthesis emerging from cultural disjuncture. As it happens, both examples involve talented individuals — Singaporeans — who were schooled in certain traditions and who were able to creatively apply their cultural resources to radically new contexts. My first example is that of Mok Wei Wei in the field of architecture. In reflecting on the sources of his creativity, he cites how his having read the Chinese classic *Shitouji* or *The Story of the Stone* (also known in English as *The Dream of the Red Chamber*) many times as a child and as an adult made an impact on his architectural design, such as recalling a particular episode in which the "precocious but brilliant" protagonist Jia Baoyu commented on garden landscape. For Mr Mok, "the detailed narration of life and rituals in the great house and its walled garden, as well as the evocation of minute movements through these spaces lodged deeply in my mind... as a student and as an architect". In time, he was influenced by contemporary architectural movements, including the work of Frank Gehry, but he also created "a rewarding path of innovation through adopting traditional and global references" (Mok, 2006).

Somewhat uncannily, *The Dream of the Red Chamber* also appears as a source of creativity for Andrew Gn, the Paris-based fashion designer, who said in a recent interview that he keeps the book by his bedside: "It has these descriptions of clothes. It was my very first way of understanding fashion." Mr Gn does not identify with Baoyu as a "weak and fragile person" but with "how he is poetic and liberal", and how "he pays attention to women, which is my obsession as well as my trade. It's all about women" (*The Straits Times*, 8 January 2007).

LONG-TERM CULTURAL CAPITAL ACCUMULATION

What, then, is the connection between a classical Chinese novel and creativity in architectural and fashion design? It is not a simple one-to-one connection. But what the two examples show is that cultural disjuncture — and, arguably, by extension, what we call "identity politics" — can be productive in the creation of new cultural forms on the part of individuals who were bred in a "traditional" environment and are thrown into new cultural environments in which they are exposed to new influences and have to critically muster their own cultural resources in creating something new. What is it about Singapore that we are not doing more of this and having more of such individuals? In thinking about the political and policy implications of the task of transforming our economic advantage into a cultural one, we do need to look at the long term. We have talked about economics and culture; now what does it mean to have cultural capital accumulation over the long term? Now, in the long term, as Keynes said, we are all dead, but if we project ten, 15, 20, 25, 30 years ahead, what is it about the culture in Singapore that makes the whole place more than economically sustainable, more than just vibrant as a tourism destination? I have suggested that cultural disjuncture — and identity politics — is not necessarily a bad thing; the kinds of tensions that accompany social pluralism and cultural diversity could lead to new cultural forms that we are unaware of right now. In highlighting the need for cultural capital accumulation in the long term, I do not intend to downplay the importance of recognising that signs of intolerance between citizens and foreigners, or among different social groups in Singapore, are likely to be rooted in a sense of inequity and a sense of insecurity among certain groups. And this is why among the kinds of identity politics that we haven't talked about is the politics of class identities, which is an understated and under-discussed research topic in Singapore.

A few more points may suffice to conclude my presentation. We also need to look at the relation between the measures that enhance security and those that promote creativity. The Singapore state is well known in making security a top priority. And it does it in such an all-encompassing and comprehensive way, whether it is related to threats such as terrorism or pandemics. But we should consider whether and how the security mentality and the plethora of security measures make an impact on the "ecology" of creativity in Singapore — in the quest for security, are the conditions for creativity in the society undermined?

Yet another point concerns the differentiation between "cultural industries" and "creative industries" — both of which contribute to the economy in terms of multiplier effects. For example, the integrated resorts (IRs) and casinos that will soon be built are, in a sense, cultural industries because gambling, like sport, is a cultural phenomenon and some may say is an intrinsic part of everyday life in Singapore or even part of "Singapore culture". So it is a huge cultural industry, which is also calculated to be profitable. But is this a *creative* industry? Does it have spin-offs for the ownership and use of intellectual property rights? Is this money that will somehow stay in the country and get rolled around, leading to new creative products? This is something that I would question. The integrated resorts will soon be built, and they will have some pride of place in Singapore's ambitions to be a vibrant global city. But in a creative city, we need a larger and more complex mix of cultural and creative industries, a kind of creative tension between the IR paradigm and other forms of creativity, perhaps coming from the edges or the periphery (e.g. from the dynamism of arts groups). In what ways are the contributions of the smaller cultural players facilitated in the longer-term cultural capital accumulation in Singapore?

Finally, among the kinds of creative capabilities that we need in Singapore is a kind of "synthetic creativity". I shall end here by quoting the words from another study on the creative city (Landry and Bianchini, 1995): "... we need a completely different type of creativity, as increasingly we know more facts but understand less. In particular, we need the creativity of being able to synthesise, to connect, to gauge impacts across different spheres of life, to see holistically, to understand how material changes affect our perceptions, to grasp the subtle ecologies of our systems of life and how to make them sustainable. We need, in other words, the skills of the broker, the person who can think across disciplines, the networker, the connector..." Singapore has been seen as a marvel of social engineering, an epitome of success in the spheres of economic development, science and technology, and governance. This could not have been achieved by relying solely on an engineering perspective or an economistic outlook without elements of "synthetic creativity" on the part of leaders, residents and guests. In the longer term, however, we need to "massively transform" our current advantages into cultural resources that will help us build a sustainable and dynamic Singapore synthesis.

REFERENCES

Benjamin, G. "The Cultural Logic of Singapore's 'Multiracialism'," in *Singapore: Society in Transition*, ed. Riaz Hassan. Oxford University Press, Singapore, 1976.

Chua, B. H. "Racial Singaporeans: Absence After the Hyphen," in *Southeast Asian Identities: Culture and the Politics of Representation in Indonesia, Malaysia, Singapore and Thailand*, ed. Joel S. Khan. Institute of Southeast Asian Studies, Singapore, 1998.

Furnivall, J. S. *Netherlands India: A Study of Plural Economy*. Macmillan, New York, 1944.

Hall, P. *Cities in Civilization*. Pantheon, New York, 1998.

Hefner, R. W. "Introduction: Multiculturalism and Citizenship in Malaysia, Singapore, and Indonesia," in *The Politics of Multiculturalism: Pluralism and Citizenship in Malaysia, Singapore, and Indonesia*, ed. Robert W. Hefner. University of Hawaii Press, Honolulu, 2001.

Kwok, K.-W. and Mariam, A. "Cultivating Citizenship and National Identity," in *Singapore: Re-engineering Success*, ed. Arun Mahizhnan and Lee Tsao Yuan. Institute of Policy Studies and Oxford University Press, Singapore, 1998.

Landry, C. and Bianchini, F. *The Creative City*. Demos, London, 1995.

Mok, W. W. *Chinese More or Less*. Kristin Feireiss, Berlin, 2006.

Tan, E. K. B. "'We, the Citizens of Singapore...': Multiethnicity, Its Evolution and Its Aberrations," in *Beyond Rituals and Riots: Ethnic Pluralism and Social Cohesion in Singapore*, ed. Lai Ah Eng. Institute of Policy Studies and Eastern University Press, Singapore, 2004.

PuruShotam, N. S. *Negotiating Language, Constructing Race: Disciplining Difference in Singapore*. Mouton de Gruyter, Berlin and New York, 1998.

The Straits Times, "Singapore Culture Unlikely to Emerge: MM," 6 September 2006.

The Straits Times, "Run(a)way Success," 8 January 2007.

CHAPTER 6

The Singaporean Artist

BEATRICE CHIA-RICHMOND

INTRODUCTION

Good morning, ladies and gentlemen. I'm Beatrice Chia-Richmond. I thought I should start my speech this morning by introducing myself. I am a director as well as an actress. I studied and trained in the United Kingdom for five years and worked for another three before returning home in 1998. These days I direct mainly for the theatre. This pretty much lumps me in the category of artists!

Usually this means that given an opportunity like this (as in given an audience of 700 people), I would be complaining, whining, criticising, and carping at the top of my lungs, trying to compress as many complaints as is possible in 20 minutes. After all, we artists are known to be an unhappy and critical lot.

I would have lashed out at the arts council on the usual bugbears — lack of funding; the MDA (Media Development Authority) for censorship policies — after all, now that we can have full frontal nudity onstage, when can we have the play that criticises the government? Then I would have railed at corporations for only importing the giant Western musicals into town; scolded the local journalists and newspapers for not understanding my work. Then I would end off triumphantly by cursing all Singaporeans who rushed off to watch West Side Story instead of a locally produced play.

I thought this would go down well! That this would definitely win me some new supporters today! But I won't do that this morning because, firstly, I've grown tired of playing the angry, cynical and frustrated artist.

That role has become too easy to play. Secondly, I've found myself in the rather ignominious position of being pretty happy with the situation we have here in Singapore, especially in the last few years.

I was asked to speak today about why I came home after a long period abroad in England, London specifically, where all the artistic and cultural action is; my reasons for staying and how I feel about being Singaporean today. I imagined, that because I am a likely candidate for the diaspora, why did I come home? What is there for me here? Or rather what is there for the Singaporean Artist that makes this place our choice over perhaps more culturally developed cities like London, Paris or New York?

I'd like at this point to shamefully confess that I was one of those Singaporeans who by the age of 15, had found everything to detest and criticise about my country. Not only were we small and insignificant, I was intensely embarrassed by our crudeness, lack of sophistication, close-mindedness… just to name a few. Stories of The Ugly Singaporean and terms like "kiasuism" with all the cartoons were never remotely endearing to me.

Brought up on a healthy diet of English literature, I was determined to leave for England. Breathe the same air as Byron and Keats. I was going to educate myself, work in England, drink my Pimms, get my British passport and never turn back.

My identity confusion, or rather, the rejection of my origins was so intense that I actually believed that I was English deep, deep down. That if you cut my veins, British blood would spurt out. Me — being the actress — had the whole accent thing down pat. I would have given Helen Mirren a run for her money. I drank at the pubs with my mates, listened to Oasis and life was "wicked, smashing, and pukka!"

One day (I was still at boarding school at this stage), I overheard a schoolmate trying to describe me to the school matron. She said, "Beatrice Chia — you know, the chinky one from Singapore, with the weird accent."

That was the first time I realised how the world would always see me first — the Chinese girl from Singapore, whether I felt it or not! Whether I accepted it or not! Your race and country are always the first things in establishing your identity. No matter how fast globalisation can blur the lines, how anonymous we can be on our blog sites, or how cosmopolitan you can be, when you come up in the flesh and blood, you are defined first by your race and country. And you never know this more acutely than when you leave your country to see yourself abroad.

In the training of an actress, or an artist for that matter, you call into examination your race; personal, cultural, traditional beliefs; and history, even more acutely. It's kind of like, you have to know who you are first before you can start portraying others. This is true for most artists from a creative and expressionist point — that an artist's first body of reference is always himself.

So, much as I had imagined myself to be Kate Winslet, I really don't think I need to tell you how many Vietnamese prostitute roles or Thai transvestite roles I was offered upon graduation in the UK.

After eight exciting years in the UK, I came home. Albeit kicking and screaming at first. Not because the Thai transvestite roles were coming in fast and furious, but because of one of the biggest things that pull at an Asian child's conscience and heart — my parents.

The first few years back were frankly an exercise in misery. The arts scene in 1998 was not thriving as it is now. Theatre was still pretty much a hobby, and at best, a long-running fringe event. A handful of professional theatre companies, shoe-string budgets, no exciting Esplanade, no nudity allowed onstage... I had one foot out of the door, passport in hand, and one foot in my paternal home. I felt very much like the unhappy child always on the brink of running away from home. Always feeling that I was missing out on some great exciting artistic opportunities in London and New York.

OPPORTUNITY FOR MISTAKES

At some point, I thought, enough carping and whining. I decided to do something here. The moment I did, I realised something invaluable Singapore has that is priceless to any artist — the opportunity to make mistakes. This is a town where you can get slaughtered as an actress, designer, director, artist, writer... but you will work again. Maybe because really we can't actually tell the good stuff from the bad. Maybe because we are so desperate for any talent, that you can mess up real bad a few times, but you will work again. At this point, I think it necessary for me to define my idea of making mistakes. By making mistakes, I mean mistakes of an artistic nature — producing work so bad that it just stinks like a rotting carcass. For example: ugly meaningless sculptures, poorly directed shows, pretentious writings, unoriginal music, awful paintings. You get the picture.

Just to prove my point further, I've brought along some of the worse reviews I've received in my career in Singapore. Here are some slaughterings of my sweat and blood. I am happy to tell you that while these reviews stung my ego like nobody's business, resulting in box-office tragedies, I survived and am still directing today.

Compare this to the ferociously critical press in London, New York or Paris. That kind of criticism is a real slaughterhouse and effectively destroys an artist's career before he can even get into second gear.

Seriously, on a less flippant note, the ability for an artist to make mistakes of this nature is a priceless one. We are a young country, with a short artistic legacy. We have no Mozarts or Michelangelos to call our own yet. The only way for us to have them is to nurture them even while they fumble and tumble through. That means taking their good stuff and bad stuff, and not shutting them down too quickly. The only way an artist grows is by simple trial and error — *error* being the operative word here today.

There is another kind of mistake that I would like to touch on — the kind of mistakes which are not box-office disasters, but where artists cut off their pubic hair or drink their own urine in public as part of performance art. (Censorship moment: I'm sorry for offending you if I have to use the word "pubic" hair. But being the responsible artist that I am, I will put my hand above my head whenever I'm going to say the word, so you can shut your ears!) In 1994, a 22-year-old artist Josef Ng cut off his [puts hand above head] pubic hair in his performance art piece in protest against the media's demonisation of 12 homosexual men. He was arrested by the police on charges of obscenity. He was eventually fined $1,000 for committing an obscene act and banned from performing in Singapore. Performance art was also banned and the ban only recently lifted.

This incident sparked off a huge uproar from the artistic community as you can imagine — because we are an unhappy reactionary lot itching for a fight. Amidst the name-calling and protestations, the role of art was once again examined, dialogue was facilitated and policies were re-examined and readjusted. The artistic leaders who could and would took a public stand. In the meantime, all of us got an education in performance art and of course the foreign media went to town ever eager to paint Singapore as the anally-retentive country run by a despot.

This event was so exciting (much more exciting than the conferring of the yearly Cultural Medallion!) that I read all about it while in England, without the Internet, in 1994. My point is this: that mistakes like that, or rather struggles like these, provoke and challenge us as a society, whether you are artistically-inclined or not. Because events like these make us think harder and question what kind of society we want. It questions censorship policies. Can we be a Renaissance City if the authorities have the ability to ban performances they deem inappropriate? Censorship debates always talk about protecting the people. But in 2007, how much of this protection is still necessary? Can we be a Global City for the arts with our censorship policies on political and sexual expression?

I see the Josef Ng episode as a useful chapter in our artistic legacy. It is often used as an iconic example during censorship debates. I don't care to make judgments on what I thought was right or wrong. That's not the point today. The point is that, these chapters make up our artistic history — which shouldn't be just an archival record of water-colour paintings, or Cultural Medallion winners. An artistic history is also made up of the struggles and dialogues between the artists and the patrons, the arts council, the audience.

ROLE OF THE SINGAPOREAN ARTIST

This is where I'd like to talk about the role of the Singaporean Artist. Because this is where I feel my role most keenly. I'm not suggesting that I too go outside the Shangri-la Ballroom and trim my [puts hand above head] pubic hair. You see, artists need a society to affect. They need a society to fill the theatres, applaud them, journalists to misunderstand them, arts council officials to pull funding from them. Because we are a young culture and policies can be re-examined and re-shaped to move with change. The Singaporean Artist plays an important role here because he can be part of that process, part of that artistic history, in his own country. He is unable to do that in London or New York, simply because he is the second-class citizen there.

SO MANY AUNTIES TO SHOCK IN SINGAPORE

My third big reason for staying in Singapore, is that there is still so much to do. Coming back from a fast and highly developed city like London or New York, where everything has been done at least 50 times over, it feels like there is nothing new to try there. Josef Ng can try trimming his [puts hand above head] pubic hair in public and the public will probably grunt and just toss him a couple of shillings. In Singapore, performance artist Vincent Leow drank his own urine in his performance art piece and caused a minor controversy. In New York in the 1970s, Japanese artist Yayoi Kusama drank her own urine and people just thought she took far too much acid.

The only thing you are trying most of the time in these fast-paced cities is just to get your foot in the door. Compare this to Singapore, where by nature of our youth, there are so many things to introduce to our people. In London's contemporary art scene, they have already progressed to Damien Hirst's rotting cow's head and sharks in formaldehyde. Here in Singapore, we haven't shocked the aunties enough! Our audiences here still go to the theatre wanting a good time. They might be hard to please at times, but essentially they still have a fresh appetite and perspective. They are not totally cynical and jaded yet, and that, to me as an artist, is priceless.

Too often we complain about our music industry, and record companies who don't believe in the Singapore musician. Our Singapore Idols cannot make it, our football team couldn't qualify for the Asian Games in Doha. Our table-tennis players are from PRC. We hope for a World Cup squad by 2010. I hope for the first Oscar for our Singaporean film makers; I'm still holding out for our version of The Beatles or The Jackson 5. I hope for the maturing of our society — the day when my country no longer needs to impose a ban on anyone or anything. And that's what is making it so exciting for me to stay — that we have so many things yet to achieve. This is what makes Singapore so attractive and so compelling for me to want to devote a large part of my life to.

Now that I've talked about the many positive aspects of staying in Singapore, I think I am entitled to some complaints.

This is a complaint for the artists: We are so bad, no, *terrible,* at taking criticism. Directors here have been known to bar reporters from reviewing their productions. This is problematic as it doesn't progress the matter of critical discourse, of which there isn't enough in Singapore.

My take on this is: Artists, defend your work; don't just leave in a huff.

To the corporate sponsors: Not everything foreign is necessarily better. Firstly as artists, I'd like to state that we have no issue with foreign talent or imported works. In fact we welcome them. But what I'm asking for is *discernment.* Know when you are getting the equivalent of the third-division team in Singapore and don't pay Premier League prices.

I am privy often to the big budgets allotted to the foreign works. Again, I have no issue with that. All I'm suggesting is, can we the local artists share some of that? Don't make us the second-class citizens to the imported works.

To the audience: Have faith in us. The good stuff is coming!

In conclusion, here are my wishes for the Singaporean Artist: Continue to complain and agitate by all means. That's what we are good at. But take responsibility for your work and the criticism you level. When criticism is levelled at the arts community in Singapore, *you* are culpable too. Don't just jump on the criticism bandwagon.

Make as many mistakes as possible in the next few years. I doubt you will get away with it in the long run.

Can we have more Josef Ngs please? Life as a Singaporean Artist has never been more exciting.

SECTION 3

New Technologies and Social Change

CHAPTER 7

New Media and the Singaporean: Rediscovering the Lost Art of Media Literacy

LIM SUN SUN

INTRODUCTION

This paper begins by discussing the different dimensions of media literacy and why it is imperative that media consumers of today possess media literacy. It highlights emerging trends in the media landscape which further challenge consumers' media literacy skills and reviews efforts which are being made to educate Singaporeans about Internet literacy. It argues that the controlled media environment in Singapore, with its traditionally paternalistic approach towards content regulation, makes Singaporeans ill-prepared for an unbridled media environment like the Internet and converged media.

THE MULTI-DIMENSIONALITY OF MEDIA LITERACY

First, let us consider the concept of media literacy. Living in a highly mediatised society, where we are informed, educated and entertained by media communications, media literacy is often assumed of consumers. Indeed, for the literate, media consumption seems to be an intuitive process where the act of reading a newspaper, watching a television programme or surfing a website can be a simple, straightforward and stimulating process. However, even for the literate, media consumption requires a wide range of skills which go beyond knowing how to read, turn on a television or use a search engine. We need skills which enable us to comprehend and evaluate the media messages which we receive. To do so effectively, media consumers need to possess both functional media literacy — knowing how to access media — and critical media literacy — being able to understand and critique media messages (Buckingham, 2005).

Media literacy has been defined as

> "a set of perspectives that we actively use to expose ourselves to the media to interpret the meaning of the messages we encounter. We build our perspectives from knowledge structures. To build our knowledge structures, we need tools and raw material. These tools are our skills. The raw material is information from the media and from the real world. Active use means that we are aware of the message and are consciously interacting with them". (Potter, 2005: 22)

The knowledge structures of media literacy comprise an understanding of media effects, media content, media industries, the real world, and the self (Potter 2005: 33). The media literate individual is someone in possession of these structures — an awareness of the effects which media can have on individuals, a realisation of why some content types are excluded while others are intensively amplified, an appreciation for who controls media content and how the political economy of the media industry is reflective of and influenced by geopolitical trends, and a sensitivity to one's own conscious and subconscious responses to media messages. This is certainly a wide range of sometimes esoteric knowledge that no average media consumer, or even media scholar, would have a complete grasp of! And yet, because of the growing importance of media in our everyday lives, there is a heightened need for the average media consumer to have such knowledge.

Our discussion on media literacy has hitherto emphasised media consumption. What of media production which is taking place through blogs, vlogs, podcasts and other online services which give individuals the ability to broadcast their opinions and share their creations? The preceding definition of media literacy and its knowledge structures would need to be augmented by a consideration of the Internet environment and the manifold affordances it provides to Internet users.

THE NEW MEDIA LANDSCAPE

The convergence of "older" media like television and print with "newer" media like the Internet and the mobile phone have exciting possibilities in both media production and media consumption. Let us consider some salient trends.

From Consumer to Producer

Media consumers today enjoy the luxury of bricolage — the ability to manipulate objects in one's milieu to incorporate ideas — due to the multifarious affordances of new media (Shih, 1998; Turkle, 1995). The digitisation of new media content and the easy availability of DIY text, image, video and audio creation/editing software facilitate the practice of bricolage. Web 2.0 refers to the growing trend of websites containing content which is generated by users and is shared on a peer-to-peer basis. Avidly embraced by consumers around the world, Web 2.0 is facilitated by file-sharing services such as Flickr (for photographs), YouTube (for videos) and Gnutella (for music files and software) which are especially popular. This popularity is due in no small part to the fact that such sites vest the individual with greater semiotic democracy and almost complete creative licence in the media content they wish to put up. Consumers enjoy the freedom to incorporate existing media into their own creations, in the process subverting conventional media and infusing it with their own ideas and values.

> "On YouTube, you can see two boys reinterpret the Pokemon theme tune, or a six-minute retelling of The Shining, this time with a happy ending, or a compact splicing-together of The Lord of the Rings and Star Wars. The creative urge goes beyond video. On Second Life, a 3-D online networking game, one member has recreated the whole of the city of Dublin for other members to see — including cafes, squares, bookshops and streets." (Bowley, 2006)

Indeed, the creative energies of media consumers have not gone unnoticed and the media industry is keen to understand audiences' creative (re)interpretations (Jenkins, 2002). Citizen journalism is another fascinating trend in media consumption. Where conventional news channels fail to cover newsworthy events, or conventional media coverage may reflect institutional biases, media consumers can restore the balance by posting their own reports of such events. Blogs, vlogs and podcasts have been avidly used by citizen journalists to share their views. The ability of individuals to publish and broadcast with very few resources expands the space for public discussion, breaking the dominance of established media institutions. Consumers today therefore have access to a wider range of perspectives, and the ability to share their own perspectives with others. In Singapore, it was found that 33 per cent of Singaporeans aged 15–49 are at least moderately involved in producing content on the Internet, such as blogs, webcasts, and podcasts (Media Development Authority, 2006a).

In the process, however, individuals who produce and broadcast media content have to be mindful that they do so without compromising on their own safety, infringing their privacy and incurring liability. When they post information about themselves on their blogs or in social networking sites, consumers need to be aware of the dangers of revealing too much personally-identifiable information, such that their own safety or privacy is compromised. When they receive public feedback on their blogs or online profiles, they have to be mentally prepared that some comments may be demoralising or malicious. When they produce content or reinterpret content created by others, they have to be conscious of protecting their own intellectual property rights while respecting others'. When they share their views online, they have to consider that these opinions will enter the public domain and may cause offence if reference is made to particular individuals or groups. Indeed, they will find themselves against public criticism which they may feel compelled to address.

From Mainstream to Niche

Rather than having mainstream, undifferentiated media content pushed towards them, today's media consumers can access niche, customised content due to the proliferation of media production and distribution platforms. With Web 2.0, consumers can access virtually limitless content through an ever expanding number of manipulable channels. They also have the option of customising the media content to suit their own tastes, preferences and lifestyles. This trend of narrow-casting is certainly an exciting one because consumers' media use is no longer constrained by the dictates of the media industry or regulatory intervention. However, it does pose an interesting challenge for parents and educators who wish to guide young people in their media consumption. When the range of media types and messages was more limited, it was somewhat easier for parents and educators to keep up with media trends. The likelihood of parents and children sharing media content was also greater when choice was more restricted. Sharing media content can play an important role in encouraging intra-family communication and intergenerational understanding (Lim, 2006). Narrowcasting and audience fragmentation may threaten such practices.

From Regulated to Unregulated

In most countries, some degree of media regulation exists. Media regulation ensures that media producers are held up to standards of providing accurate and reliable content which does not offend segments of the society and create schisms. In Singapore, parents have been able to rely on official state censorship to ensure that their children are not exposed to offensive or age-inappropriate content. With the advent of the Internet, however, its unregulated nature means that parents have to be more pro-active and vigilant about guiding their children's media use. Online information is so copious and of such varied provenance that both adults and children alike need critical media literacy skills to sieve out less credible information.

Web forums are an extremely good example of an online platform that demands critical media literacy of its users. Discussions on web forums are largely unmoderated except in such circumstances as anti-social posts, flaming, or posts which contravene the aims and objectives of the forum. A quick survey of discussion forums frequented by young Singaporeans such as flowerpod.com or hardwarezone.com will reveal that many teens use these forums to solicit information on a wide variety of topics, for example, the best schools, product reviews, and restaurant tips. On such topics, the exchange is often lively and many practical hints can be gleaned. However, sensitive subjects such as sexuality invite posts which can misinform and mislead young people, with potentially drastic consequences. For example, posts providing details of illegal abortion clinics and erroneous information on the spread of sexually transmitted diseases may be seen on such forums. Given the awkwardness which young people experience in broaching topics such as sexuality with their parents or teachers, their inclination to use discussion forums to clear their doubts may be strong. In which case, it is all the more important that the young, uninitiated Internet user is able to critically evaluate the veracity of different online sources.

From Reception to Transaction

As opposed to merely receiving information via different media channels, new media services provide a wealth of opportunities for consumers to engage in transactions. They can sell, purchase and trade items, manage their bank accounts, utilise e-government services etc. Such services liberate consumers from geographical and temporal constraints as they can access an extensive range of international services at a personally convenient time. However, as more online entities conduct dataveillance of consumers — the "systematic use of personal data systems in the investigation or monitoring of the actions or communications of one or more persons" (Clarke, 1988) — consumer concerns about privacy are set to grow. The improper acquisition, storage, use and transmission of consumers' personal information by online entities may lead consumers to become the targets of spam, lose sensitive personal information, and perhaps even fall prey to identity theft (Wang *et al.*, 1998). Unless consumers know how to engage in self-protective behaviour such as opting out, using privacy-enhancing technologies, reading privacy policies, or checking trust marks, they are vulnerable to such security risks and invasions of privacy (Sheehan and Hoy, 1999).

From Communication to Networking

We have long been able to communicate through mediated channels, but the scale of communication seen today is unprecedented, allowing individuals to build extensive social networks that transcend social and geographical barriers. Broadening one's circle of friends and acquaintances via social networking services like Friendster, MySpace, Xanga, Multiply and Who Lives Near You is becoming more common amongst youths. These services allow users to profile themselves by putting up photographs, noting their demographic information and interests, charting the breadth of their social circles and sharing blogs, music and videos. There is also space for other members to leave messages for or about you and these messages can be read by other members. Some services like Who Lives Near You have slightly different features, which are able to identify for users which of the site's other users live in their neighbourhoods. In general though, these sites offer similar functions and are premised on the assumption that people seek to grow their social networks through connections with friends and friends of friends and so on. Friends can also post testimonials about each other. Relationships forged on such networks can remain at a superficial level or be deepened at the initiative of the individuals concerned. Interactions may be either online only, or complement and/or be supplemented by offline face-to-face interaction.

The use of social networking sites for expanding one's social network has obvious advantages. Users have great flexibility in impression management — enjoying the luxury of time to consider how they wish to present themselves to the world so that they can achieve desirable outcomes (O'Sullivan, 2000). The pressures of socialising face-to-face, with its risks of rejection, are less pronounced online. On the other hand, an online record of the wealth or paucity of one's social network can also affect one's public image and sense of self-esteem. Youths may therefore feel pressured to ensure that their social networks appear extensive and include the "right" people. Of course, other dangers such as the possibility of being "groomed" by online predators also exist on such sites.

Opportunities and Risks

Clearly, with the shift towards greater autonomy, increased variety of communication content and channels for media consumers, there are benefits to be enjoyed but also inherent risks.

Opportunities	Risks
• Access to limitless, customisable, niche content • Ease of content creation and distribution • Social networking • Identity exploration • Role experimentation	• Privacy risks — identity theft, spyware • Security risks — malware, sexual predation • Social psychological effects — disinhibition, deindividuation • Legal — intellectual property infringement, defamation, sedition

Rather than descend into a moral panic about potential risks, a common knee-jerk reaction of parents and educators, we should find ways of coping with these risks so that we can capitalise on the opportunities. To this end, we need to actively foster critical Internet literacy so that consumers can use new media services strategically, minimising harm and maximising benefits.

INTERNET LITERACY EDUCATION IN SINGAPORE

Efforts have been made by regulatory bodies like the Infocomm Development Authority (IDA) and the Media Development Authority (MDA) to step up Internet literacy in Singapore.

The IDA has conducted many important and useful surveys which further our understanding of the state of infocomm adoption in Singapore. Its Annual Survey on Infocomm Usage in Households and by Individuals (Infocomm Development Authority of Singapore, 2006) provide concrete data for educators and policy makers to better formulate infocomm access and Internet literacy programmes. The latest survey shows that Singapore has one of the highest mobile-phone penetration rates in the world at 76 per cent, and the proportion of households with home access to a computer is 74 per cent, a figure which has remained unchanged since 2003. Of those households without home computer access, "no necessity" (35 per cent) and "lack of skills" (32 per cent) were the top two reasons cited for not having access to a computer at home. Clearly, such households recognise that without basic infocomm literacy, having access to computers may be futile. As for Internet access, the proportion of households with access to the Internet at home was 66 per cent. Amongst households with computer access but not Internet access, 29 per cent cited "access costs are too high" as the top main reason for not getting access, while 10 per cent cited "concern about exposure to inappropriate/harmful content" and 6 per cent cited "lack of skills". Households in the latter two groups would be appropriate targets of Internet literacy programmes. Interestingly, only 1 per cent cited "privacy concerns" and another 1 per cent, "security concerns", suggesting that these two factors are not strong deterrents to Internet adoption in Singapore.

Their surveys also go further into exploring consumer awareness of Internet security. The IDA found that 54 per cent of home Internet users aged 15 and above have experienced a virus attack before, while 81 per cent protect their computer with virus-checking software, 56 per cent with anti-spyware software and 53 per cent with firewalls. Armed with such data, the IDA has in place a National Infocomm Security Awareness Programme that strives to educate consumers about infocomm security and increase their awareness of this issue. This programme employs online and print media advertorials, web portals, road shows, seminars and educational learning resources to foster infocomm security awareness amongst Internet users. Of late, the IDA has also launched the Infocomm Club programme in schools to equip young people with the skills to use IT in an enjoyable and productive fashion.

For its part, the MDA has introduced the Cyber Wellness concept which comprises four core values (see box). The MDA works with public and private partners to run Cyber Wellness programmes that promote responsible Internet use. As of 2006, these programmes have reached out to more than 330,000 people. Notably, the MDA-supported Cyberspace Risks and where U Seek Help (CRuSH) programme run by Touch Community Services since 2005 has taught more than 55,000 students on the importance of safe surfing through road shows and mentoring and training sessions.

Media Development Authority's Cyber Wellness Programme

Cyber Wellness refers to the positive well-being of Internet users. It involves raising the awareness of Singaporeans on the positive effects of the Internet, the risks of harmful online behaviour, and how one can protect oneself and other Internet users from such undesired behaviour. The four core values of Cyber Wellness are:

1. Balanced Lifestyle

Users should embrace the Internet and integrate it into their daily lives. However, a balance needs to be maintained between the physical and virtual world.

2. Embracing the Net and Inspiring Others

Users should harness the positive powers of the Internet to be proactive contributors who can inspire and benefit others.

3. Astuteness

Users need to be astute and street-smart when navigating the Internet including disseminating information online and befriending strangers on social sites.

4. Respect and Responsibility

Users need to have a sense of respect for the Internet and for other individuals. They should not abuse the power of the Internet nor condone subversive content.

SOURCE Media Development Authority.

The MDA has also introduced MediAction!, an annual series of events geared towards equipping Singaporeans with the skills to consume and produce media content (Media Development Authority, 2006a). 2006 also saw the formation of the National Internet Advisory Committee (NIAC)-Community Advisory Committee (CAC) to focus on "cultivating a media literate population" (Media Development Authority, 2006b). Subsumed under this committee was the Parents Advisory Group for the Internet

(PAGi) which had actively conducted talks and courses educating the public on effective parental supervision of Internet use.

Instructing Parents, Teaching Children

This begs the question of how critical Internet literacy can be imparted, especially to youths. Are parents necessarily the most well-placed individuals to inculcate media literacy skills in their children? A discernible thread in the communications literature identifies the challenges which Asian parents face in exercising supervision over their children's new media usage. See, for example, the situation in Indonesia (Guntarto, 2001), Thailand (Komoselvin, 2002), Vietnam (Nguyen and Nguyen, 2002) and Singapore (Lim and Tan, 2004). There is often a perceptible gap between children's and parents' knowledge of new media and infocomm technologies, arising from the tendency for young people to appropriate these new media to a much higher degree than their parents do. This impacts negatively on the parents' traditional roles as teachers. With older media, this issue was less pressing as it is comparatively easier for parents to oversee activities like television viewing which do not require special technical knowledge.

In this regard, it is imperative that we understand the contexts of youths' day-to-day media consumption so that critical literacy skills can be effectively imparted to youths. In order for youths to benefit from new media, it is imperative that the pace of innovation be matched by the rate of acquisition of critical media literacy skills. Urban youths who tend to access new media in the home or school environment should be targeted by media literacy training programmes for parents and youths. Indeed, parents can play a key role in fostering critical media literacy. They must fight the urge to engage in parental disintermediation, where they resign themselves to the fact that they know less about media and technology than their children and refrain from supervising their child's media use (Turow and Nir, 2003). To do so would be a mistake, for even if parents know less about technology than their children do, their wealth of life experiences and real world knowledge give them some ability to evaluate media messages critically. Of course, this ability would be enhanced if the parents seek to actively acquire media literacy skills of their own. Arguably, Singaporeans face the additional challenge of having had their media literacy skills atrophy through very efficient content regulation by the government. Singaporean parents have been able to leave their children largely unsupervised when watching television or reading newspapers and magazines, knowing full well that any violence, nudity, coarse language and extremist views would already have been filtered out by efficient government censorship. With the domination of Singapore's media landscape by two well-respected government-linked companies, Singapore Press Holdings and MediaCorp, consumers have also come to expect reliable and accurate information from these two sources.

With the Internet unleashing unregulated content into the hitherto sheltered Singaporean household, we must pause to consider whether our media literacy skills are adequately suited to cope with this new media landscape. There is perhaps some room for optimism as a recent MDA survey revealed that 55 per cent of Singaporeans aged 15–49 believed that they were at least moderately able to analyse and discern information on the Internet (Media Development Authority, 2006c). It should be noted, however, that this is a self-assessment and not an objective test administered by a third party. In-depth qualitative studies need to be conducted to better understand where and how Singaporeans' media literacy skills are lacking.

CONCLUSION

Agencies like the MDA, the IDA and the Ministry of Education must press on in their efforts to inculcate infocomm literacy skills, and also shift the focus to the fostering of critical media literacy skills. At the same time, however, the onus is also on the individual to be alert to new media trends and apprise themselves of opportunities and risks. For in the age of the media prosumer — a consumer and producer of media — media affordances come with both possibilities and responsibilities.

REFERENCES

Bowley, G. "The High Priestess of Internet Friendship," *Financial Times,* 27 October 2006.

Buckingham, D. "The Media Literacy of Children and Young People: A Review of the Research Literature," Ofcom, London, 2005.

Clarke, R. "Information Technology and Dataveillance," *Communications of the ACM*, Vol. 31, No. 5 (1988): 498–512.

Guntarto, B. "Internet and the New Media: Challenge for Indonesian Children," *Media Asia*, Vol. 28, No. 4 (2001): 195–203.

Infocomm Development Authority of Singapore. "Annual Survey on Infocomm Usage in Households and By Individuals for 2005," 2006. Retrieved 30 December 2006 from http://www.ida.gov.sg/Publications/20061207182001.aspx.

Jenkins, H. "Interactive Audiences," in *The New Media Book*, ed. Harries, D. British Film Institute, London, 2002.

Komoselvin, R. "Education, Encouragement, Self-regulation: Children and the Internet in Thailand," in *Kids On-Line: Promoting Responsible Use and a Safe Environment on the Net in Asia*, ed. Shetty, K. AMIC/NTU, Singapore, 2002.

Lim, S. S. "From Cultural to Information Revolution: ICT Domestication by Middle-class Families in Urban China," in *Domestication of Media and Technology*, ed. Hartmann, M., Berker, T., Punie, Y. and Ward, K. Open University Press, Maidenhead, 2006.

Lim, S. S. and Tan, Y. L. "Parental Control of New Media Usage in Singapore — The Challenges of Infocomm Illiteracy," *Australian Journal of Communication*, Vol. 31, No. 1 (2004): 57–74.

Media Development Authority (2006a). "MDA Encourages Singaporeans to Take Action with MEDIAction!" Retrieved from http://www.mda.gov.sg/wms.www/actualTransferrer.aspx?c=2.1.29.&sid=738&eid=-1&fid=-1.

Media Development Authority (2006b). "Community Advisory Committee Formed to Focus on Media Literacy." Retrieved from http://www.mda.gov.sg/wms.www/actualTransferrer.aspx?c=2.1.55.&sid=704&eid=-1&fid=-1.

Media Development Authority (2006c). "Results from Survey Show 65 Per Cent of Singaporeans Use Internet in Their Daily Lives." Retrieved from http://www.mda.gov.sg/wms.www/thenewsdesk.aspx?sid=753.

Nguyen, Q. N. and Nguyen, Q. T. "An Exploratory Study of Children and Electronic Games in Vietnam," in *Kids On-Line: Promoting Responsible Use and a Safe Environment on the Net in Asia*, ed. Shetty, K. AMIC/NTU, Singapore, 2002.

O'Sullivan, P. "What You Don't Know Won't Hurt Me: Impression Management Functions of Communication Channels in Relationships," *Human Communication Research*, Vol. 26 (2000): 403–431.

Potter, W. J. *Media Literacy*. Sage, Thousand Oaks, CA, 2005.

Sheehan, K. B. and Hoy M. G. "Flaming, Complaining, Abstaining: How Online Users Respond to Privacy Concerns," *Journal of Advertising*, Vol. 28, No. 3 (1999): 36–51.

Shih, E. C. F. "Conceptualising Consumer Experiences in Cyberspace," *European Journal of Marketing*, Vol. 32 No. 7/8 (1998): 655–663.

Turkle, S. *Life On The Screen*. Simon and Schuster, New York, 1995.

Turow, J. and Nir, L. "The Internet and the Family: The Views of Parents and Youngsters," in *The Wired Homestead — an MIT Press Sourcebook on the Internet and the Family*, ed. Turow, J. and Kavanaugh, A. MIT Press, Cambridge, MA., 2003.

Wang, H. Q. *et al.* "Consumer Privacy Concerns about Internet Marketing," *Communications of the ACM*, Vol. 41, No. 3 (1998): 63–70.

CHAPTER 8

Singapore's Emerging Informal Public Sphere

CHERIAN GEORGE

INTRODUCTION

One of the major media milestones that Singapore crossed in 2006 was the introduction of High Definition TV or HDTV. Offering extremely high resolution widescreen pictures coupled with the benefits of digital interactivity and more channels, HDTV was showed off at selected public places such as community clubs. The Media Development Authority and private-sector technology partners were banking on the likelihood that crystal clear images of glistening dew on the tip of a leaf in a nature documentary, or the instantly spottable golf ball sent zipping through the air by Tiger Woods, would convince Singaporeans to embrace HDTV and justify the investments in this new technology.

Ironically, however, the televisual trend that caught on in 2006 without any formal encouragement had a screen size of about one-fortieth of a HDTV set, and grainy pictures reminiscent of the work of an impressionist painter not wearing his glasses. This was of course YouTube, which became the flagship of the movement known as user-generated content — a movement supposedly so significant that *Time* magazine named You as its Person of the Year. Together with blogs and other communication technologies, YouTube is helping to turn "the journalism of the lecture" into "the journalism of the conversation" (Gillmor, 2004).

One key question is what impact this development has for the public sphere — the space where citizens discuss and deliberate matters of common interest and public concern, and hold the state accountable. It is tempting to frame the trend in competitive terms, as a battle between traditional mainstream media and alternative new media for supremacy in the public sphere. However, although there are some competitive aspects to this dynamic, it is useful to view it also as a complementary relationship, just as "Off-Off-Broadway" productions are part of the same ecology that also produces high-end Broadway hits, and garage tinkering is organically tied to high-end research laboratories. Social theorists suggest that the public sphere isn't and shouldn't be unitary or monolithic. We should instead think in terms of a *formal* public sphere that is complemented by multiple *informal* public spheres. The formal public sphere is where the broadest national issues are discussed, consensus is sought, and negotiation and social conciliation is practised. It's where people figure out their common interests and work through their shared problems as a Public. This is the role that national newspapers and broadcast channels are well suited for.[1] Indeed, to the extent that nations are imagined communities (Anderson, 1991), the national media are principal "imagineers", to borrow a job title from the Disney corporation.

However, the norms and protocols that are necessary for the proper functioning of the formal public sphere typically exclude and marginalise minority points of view, even in the freest of liberal societies. Therefore, the informal public sphere plays an important role in allowing broad participation; they are the spaces where people can share ideas more freely. "Here," says Jürgen Habermas (1996), "new problem situations can be perceived more sensitively, discourses aimed at achieving self-understanding can be conducted more widely and expressively, collective identities and need interpretations can be articulated with fewer compulsions than is the case in procedurally regulated [formal] public spheres." These are the roles that user-generated alternative media, with their low barriers to entry, are exceedingly well suited to.

An informal public sphere is not new either as an idea or as a social phenomenon. In Singapore, however, it may take some getting used to. People's Action Party (PAP) ideology has emphasised consensus rather than the expression of dissonant viewpoints. This ideology has been institutionalised in the media system, with strict licensing laws ensuring that the mainstream media are under the duopoly control of large, trusted corporations. The proliferation of niche and alternative media has put pressure on mainstream media and on the government.

CHALLENGES FOR MAINSTREAM MEDIA

Singapore's mainstream media are being challenged on a number of fronts — profits, readership and viewership, and influence. Blogs and other user-generated content are only part of that challenge, and indeed the mainstream media were in gradual decline long before blogging. The audience's attention is dissipating across a wider diversity of media forms. At the same time, the advertisers that used to reward newspapers for their ability to congregate the masses now have alternative outlets, ranging from niche magazines to public transport vehicles.

This decline needs to be put in perspective: newspapers are still the most profitable media businesses, and still occupy the commanding heights of the news business; it is just that its degree of dominance is slipping.[2] Mainstream media's superior resources should mean that they will continue to be able to offer more and better content than most of their competitors. However, as general interest media, the mainstream media cannot hope to serve all of the people all of the time. As Singapore society becomes increasingly complex and variegated, as sub-cultures proliferate, and as tastes become increasingly specialised, it is getting tougher for the national media to serve all of the people even *some* of the time.

Media companies around the world are responding by spinning off more niche publications and supplements. There are two problems with this approach. One is that not all readers are created equal in media companies' eyes. If you have the disposable income to shop for cars, luxury watches, designer clothes and spa vacations, media companies will pander to you in order to deliver you to their advertisers. Readers of lesser means are less attractive to advertisers and are thus unlikely to see the creation of magazines, supplements or special sections on such themes as how to reduce household bills or maintain emotional health through methods other than shopping. In a country with a growing and already sizeable socio-economic divide, there is a risk of large segments being unserved by the media. There is another problem with going niche. While people want to nurture their own unique identities and pursue their own interests and lifestyles, society as a whole would be poorer if there were no common spaces left. If the national media appealed to all of the people *none* of the time, one would have to ask if there is anything Singaporean about Singapore any more. Therefore, the mainstream media need to balance individual desires for niche content with the social need for common spaces. This is easier said than done, but must remain a top priority.

Space for Alternative Views

Another challenge faced by mainstream media is their handicap in reflecting alternative views. This is the result of two distinct attributes. The most obvious is the burden of operating under a government licence. The regulatory regime requires mainstream media not to try to set the political agenda, which in practice means that editors are expected to filter out or at least not over-amplify views that contradict government positions on key principles or policies. Alternative media on the Internet are not subject to discretionary licensing and therefore enjoy much wider latitude in expressing contrary views (George, 2006).

In addition to political constraints in countries such as Singapore, the mainstream media around the world also operate with a technical disadvantage. Paradoxically, the professional operations and high production values associated with mainstream media seem to be creating a counter-demand for a more personal, supposedly authentic experience via cottage-industry media. This phenomenon is not unique to the news media industry: it seems to apply to most cultural and lifestyle products (Carroll and Hannan, 2000). Thus, there are beer connoisseurs who would shun Tiger and Heineken and opt for microbrews and homebrews, despite the latter's inconsistent quality. Similarly, music lovers may scoff at assembly-line boy bands, no matter how slick, and seek out underground, garage bands. This tendency may also explain the aforementioned appeal of YouTube, despite the seemingly superior quality control exercised by the TV industry. The imperfect but personally crafted and authentic is being embraced as an antidote to the impersonal and industrial, no matter how professional the latter.

Can Singapore's mainstream media overcome this twin handicap of licensing and industrial standards? The dichotomous regulatory regime — with stricter supervision of mainstream media and more latitude for niche and/or alternative media — is likely to be preserved. However, since think tanks are supposed to think the unthinkable, I would be shortchanging this IPS forum if I failed to at least raise the question of reviewing the Newspaper and Printing Presses Act. It is noteworthy that in Malaysia, which has a comparable newspaper permit system, the Malaysian Human Rights Commission has called for the following amendments: making permits permanent rather than requiring annual renewal; making the granting of permits automatic, subject to objections from security agencies; and requiring the government to publish reasons for permit rejection, which can then be challenged in court (Suhakam, 2003).

Liberalising licensing rules (and aggressively upholding competition laws) could have the positive effect of diversifying the regulated mainstream segment of Singapore's media. Media entrepreneurs and professionals would have the freedom to explore new business models and editorial concepts. While these are unlikely to displace *The Straits Times* as the country's number one daily, they could offer the public options that do not currently exist. In other sectors such as education and healthcare, industry restructuring over the past decade has allowed the emergence of more providers and allowed them greater autonomy, thus multiplying choice for Singaporeans. Regulators of local news media are relative laggards in this regard.

Realistically speaking, Singapore is unlikely to engage the question of licensing any time in the near future; it may be more practical to consider less out-of-the-box options. Even if the letter of the law is not revised, the government needs to adapt to a changing environment and calibrate its controls accordingly. In supervising the mainstream media, regulators and internal gatekeepers should avoid widening the gap between mainstream and alternative media. I hesitate to call it a *credibility* gap, because most people do believe that the mainstream media are by and large accurate and believable. For reasons I have touched on earlier, it should perhaps be called an *authenticity* gap — the mainstream media are seen as somehow failing to provide an authentic experience; to be presenting the news accurately, yes, but not *for* you and me — unlike, say, a favourite blog.

Mainstream media can try to respond by providing more space for user-generated content and a sampling of that other world, which is precisely what *The Straits Times* is trying to do through STOMP and what *Today* tried to do by enlisting the blogger, mr brown, as a columnist. The failure of that experiment and its backfiring on *Today*'s reputation showed how dicey this challenge is. Responding to one of mr brown's *Today* columns, the government said that while mr brown was entitled to his views, "opinions which are widely circulated in a regular column in a serious newspaper should meet higher standards" (Bhavani, 2006). It added, "If a columnist presents himself as a non-political observer, while exploiting his access to the mass media to undermine the Government's standing with the electorate, then he is no longer a constructive critic, but a partisan player in politics."

Bridging the Divide

It is unclear whether *Today*'s immediate termination of mr brown's column was instigated by the government. Undoubtedly, though, the authorities believed that the particular offending article should not have been published in that form. Mainstream media editors have thus been sternly reminded not to abdicate their responsibility, as gatekeepers of the formal public sphere, to filter the strident voices and other noise of the hoi polloi. For the mainstream media's own good as well as for Singapore's, however, we should avoid erecting a firewall between mainstream and alternative media. Ideas need to flow between the two. The national media should have the latitude to reflect the buzz of alternative spaces. But, after the government's statements in 2006, can they? The sternness of their warning notwithstanding, the authorities may not be totally opposed to newspapers reporting or republishing online viewpoints as long as three criteria are met. First, of course, the statements quoted must not cross any boundaries of law or good taste. Second, avant-garde or minority views should not be misrepresented as reflecting mainstream or majority views. Third, the mainstream media should be mindful of the power they possess to bequeath symbolic status on the people and perspectives they give space to, and should therefore be judicious in whether and how they do so.

These may seem onerous rules, but they are not impossible to work with. Existing journalistic conventions allow newspapers to carry diverse content by applying a range of editing standards, which are signalled clearly to the reader. For example, regular readers know that the views that *The Straits Times* regards as most authoritative are to be found in its own editorial and in columns such as "Thinking Aloud". At the other extreme are its user-generated content pages — and even among these there is a clear hierarchy, with the "Forum" page at the top and other sections for reader contributions — including online views — given lower status. Similarly, clear signalling tells the reader that the "YouthInk" is not to be treated as seriously as more grown-up columns. The issue is not so much that readers are likely to get confused, but that editors require deniability: for their own protection, in a tightly regulated environment, they need to be able to distance themselves from content that they carry for the sake of providing a comprehensive range of viewpoints. In hindsight, perhaps *Today*'s mistake was to give mr brown's column the same look and feel of its more elevated columns, thus apparently giving the editors' stamp of approval to the arguments therein. *Today*'s relatively small staff of full-time writers creates a greater reliance on user-generated content; these and even humour columns are not distinguished particularly clearly from more considered viewpoints. To borrow the words of Singapore's eloquent former information minister George Yeo, *Today*'s design was and continues to be a case of *boh tua boh suay*.[3]

All in all, mainstream media editors can probably be trusted to preserve the distinction between formal and informal public spheres, and not to go overboard with user-generated content. After all, it would be self-defeating to do so, compromising their main competitive advantage in professionally produced content. However, there is a real risk that certain other professional standards will be compromised due to the competitive pressure posed by alternative media. Digital delivery and fewer layers of checks sometimes enable alternative media to be the first with the news. Professional journalists know that they are supposed to "get it first but first get it right". Unfortunately, once the alternative media release a piece of news, there is pressure on mainstream media to publish it on the grounds that it is already "out there". There is plenty of evidence worldwide to suggest that this risk is already materialising, short-circuiting the standard, rigorous checks that journalists know they are supposed to exercise (Kovach and Rosenstiel, 1999). Usually, newspapers will try to hide their less scrupulous judgments with a fig leaf, suggesting that although the gossip they are recirculating has not been verified, the fact that is creating a buzz is eminently newsworthy and reportable. Singapore's national newspaper is not immune to such tendencies: the front page of *The Sunday Times* was recently splashed with sexy photos of a model that, according to online speculation, was the Mongolian woman who had been murdered in Malaysia. It turned out that she was a Korean model unconnected with the sordid affair.

In appealing to the mainstream media not to imitate the alternative media in some respects, I do not want to give the impression that the national newspapers and broadcasters are always the paragons of virtue and guardians of high standards, while the alternative media are irresponsible and anti-national. On the contrary, with mainstream media becoming increasingly commercial in its impulses, the informal public sphere is seen by many Singaporeans as the more hospitable space for contributing to public life. Indeed, one could say that there is at least as much nation-building going on in the alternative media as there is in the national mainstream media. Of course, if you define nation-building in old-fashioned top-down terms — equating it merely with treating the nation's leaders with deference and amplifying their messages — then the mainstream media have the edge. However, if we adopt the contemporary understanding of nation-building as a bottom-up process of active citizenship, *à la* Singapore 21 and Remaking Singapore, then the action is increasingly in the alternative media. In a growing number of sectors — heritage and history, the arts, natural history and the environment, local music and culture, even the National Service experience — the most passionate and knowledgeable efforts to connect Singaporeans with their nation are taking place in the informal public sphere.

Increasingly, the national media are adopting commercial marketability rather than nation-building as their touchstone. They are getting away with it partly because they are careful to continue playing their traditional top-down nation-building role and thus appease their political masters. Besides, they are business entities, it's their money, and it's their prerogative to make investment decisions. On the other hand, Singapore's media giants are protected by government licensing. As custodians of scarce, publicly granted publishing and broadcasting permits, they owe a fiduciary responsibility to the public. Furthermore, if the news media choose to be ever more entertainment-driven, consumer-driven and accommodating to advertisers, then the traditional professional values of journalism as a public service will be increasingly marginalised. This is a worldwide trend, prompting the *Economist* (2006) to speculate that the mission of high-quality journalism will have to find a new home, migrating from newspapers to other types of organisation, such as NGOs and citizen groups.

CHALLENGES FOR GOVERNMENT

Singapore's emerging informal public sphere poses special challenges for a government accustomed to dealing with the public through the more pliable and predictable national media. One valid concern is whether public communication during national crises will be compromised by rumour and disinformation through alternative media. This risk may be overblown. Particularly during emergencies, people seek reliable and credible sources. As long as the government's public communication is prompt, comprehensive, transparent, and receives full mainstream media support, false information will gain no foothold. For example, public communication during the Sars epidemic and after the Jemaah Islamiyah (JI) arrests was widely viewed as a success, despite the existence of unregulated Internet media. It is when there is a lack of reliable information that rumour will fill the vacuum. If the government learns from positive examples such as its management of Sars and the Jemaah Islamiyah (JI) affair, there is little reason to fear the blogosphere during emergencies.

Aside from emergencies, it is certainly the case that government efforts to explain its policies must now contend with a counter-discourse from alternative media. The record so far would have probably convinced the government that much of this counter-discourse is of low quality and not particularly productive. On the other hand, the government appears to have correctly assessed that it is neither possible nor necessary to extend its "nail every lie" approach into the alternative online space. The noise on the blogosphere may be irritating to policy-makers but there is little evidence that it has corrupted the public mind or compromised the quality of governance. In most policy areas, the critical question is not how much noise surrounds the public, but how the public ultimately acts at the point of decision. Arguably, the public has shown itself to be capable and willing to act on sound information rather than over-react to every unreliable morsel that they come across online or in coffeeshop talk.

Probably the hardest development for the government to get used to will be the idea that it can no longer monopolise the shaping of the broad public agenda and the managing of its own reputation and status. We are witnessing a democratisation of influence. In the alternative media, reputation matters but rank and status *per se* are not respected. In engaging online discourse, the government cannot "pull rank". Arguments will be won or lost on their merits.

I have tried to argue here that alternative media have contributed to a reconstitution of the public sphere in Singapore. In addition to the formal public sphere — where national-level policy debates take place and where the public decides on its common future — there is a growing informal public sphere, with lower barriers to entry admitting a greater diversity of views and with a greater tolerance for inadequately processed information. Most of the discussion in the mainstream media, dominated by politicians and mainstream commentators, has predictably focused on the negative implications of this widening informal public sphere. However, there are a number of possible positive outcomes. First, wider and more intense deliberation can lead to more sensitive and nuanced policy-making. This is not automatic, but would depend on the aggregation, organisation and translation of myriad points of view into forms that can be acted on. Such mediation can be done by individuals and groups within government or the mainstream media or the people sector — and preferably all of them. Second, unruly communication in the informal public sphere may seem unconstructive, but public servants need greater exposure to such unregulated interventions if only for their professional development. The requisite skills for dealing with such challenges may have been irrelevant for technocratic "head versus heart" decision-making, but they are increasingly important for a new kind of politics in which intangibles such as values, culture and identity have come to the fore. Third, the informal public sphere has a nation-building function. This is not only because active participants feel a greater sense of ownership in the nation, but also because of the positive externalities that voluntary participation generates. The small communities of common interest that are coalescing around blogs and other alternative media can be building blocks for nation-building.

The communication that takes place in the informal public sphere has many obvious flaws. A lot of what circulates is silly and irrelevant at best, and at worst, tasteless, mean, small-hearted, intolerant and stupid. One hopes that 2007 and the coming years will see a certain maturing of the alternative media. This, however, would pose an intriguing dilemma for the establishment. As things stand, the superior quality of the mainstream national media benefits established political and business interests: the media are used as a vehicle of influence by the government, and as a wellspring of profit by big business. The media of the informal public sphere are not so easily harnessed. This matters little at present, because of the quality divide between mainstream and alternative media. However, if the divide were to disappear, the alternative media would generate its own quality dividend — which would not automatically accrue to the political and commercial powers that be, but would instead be up for grabs by new players. While Singapore's political and media elite routinely rail against the inferiority of blogs and other informal media, there is only one thing worse for the elite: alternative media that actually do improve in quality, reach and influence.

ENDNOTES

1. Whether the mainstream media actually do fulfil the role of the public sphere is debatable. Habermas, who is most associated with this idea, observed that the media are dominated by powerful political and economic interests that subvert their potential as agents of the public sphere.

2. Singapore Press Holdings' core operations showed net earnings of $361.1 million for the year ended 31 August 2006. In line with an improving economy, this represented 2.6 per cent growth over the previous year.

3. In Hokkein, literally means "no big, no small".

REFERENCES

Anderson, B. *Imagined Communities: Reflections on the Origins and Spread of Nationalism.* Verso, London, 1991.

Bhavani, K. "Distorting the Truth, mr brown?" *Today*, 3 July 2006.

Carroll, G. and Hannan M. *The Demography of Corporations and Industries.* Princeton University Press, Princeton, New Jersey, 2000.

Economist. "Who Killed the Newspaper?" 24 August 2006.

George, C. *Contentious Journalism: Towards Democratic Discourse in Malaysia and Singapore.* Singapore University Press, Singapore, and University of Washington Press, Seattle, Washington, 2006.

Gillmor, D. *We the Media.* O'Reilly Media, Sebastopol, CA, 2004.

Habermas, J. *Between Facts and Norms.* MIT Press, Cambridge, MA, 1996.

Kovach, B. and Rosenstiel T. *Warp Speed: America in the Age of Mixed Media.* Century Foundation, New York, 1999.

Suhakam. *A Case for Media Freedom: Report of Suhakam's Workshop on Freedom of the Media.* Human Rights Commission of Malaysia, Kuala Lumpur, Malaysia, 2003. Available online at http://www.suhakam.org.my.

SECTION 4

A New ASEAN

CHAPTER 9

ASEAN at 40: Ascendant or Decadent?

AMITAV ACHARYA

ASEAN celebrates 40 years of its existence this year. Let me try to give you an overview of what ASEAN has accomplished and what it needs to accomplish in order to remain relevant.

ACCOMPLISHMENTS

Survival Despite Skepticism

ASEAN has seen four major successes. The first is its survival despite skepticism from many. Now if you look at ASEAN's founding in 1967 or even the first five years or so of ASEAN's existence, you won't find too many people outside of Southeast Asia — from the former colonial powers like Britain, to Western academics or governments such as the United States — who were sort of optimistic that ASEAN would survive. I have checked the diplomatic documents.

By the way, if you are into history, this is the year that under the 30-year rule, British Foreign Service documents in the British archives pertaining to the founding of ASEAN will be released. The released documents in them record how ASEAN had progressed from 1967 to 1976. You will find a lot of pessimism.

There was good reason for that because a couple of attempts at regional organisation had failed. There was the Association of Southeast Asia. Then there was Maphilindo — Malaysia, the Philippines, and Indonesia. None of them lasted. And it was not just Southeast Asia which had not been able to successfully create a regional institution until ASEAN. Many parts of the so-called developing world, or the Third World, also had a lot of difficulties creating successful regional institutions. The success was in Europe, with the European Economic Community, which is now the European Union.

So most people were very skeptical about whether ASEAN would survive. Of course, to make matters worse, there was the Sabah dispute between Malaysia and the Philippines. That almost led the two countries to go to war. ASEAN itself came about partly because of the reconciliation and rapprochement between Indonesia and Malaysia following *Konfrontasi*. But relations were still a bit tense. In particular, relations between Singapore and Indonesia were very tense over the execution of the two marines by Singapore. So nobody was sure that ASEAN would last 10 years.

But the skepticism has proven to be unfounded. ASEAN has survived for 40 years, and it has flourished. It has gone on to different stages of course. That is a great achievement.

But more than that, ASEAN has been successful in avoiding war amongst its members. Now this is also very rare in the history of international or regional organisations. Of course, Vietnam intervened in Cambodia in 1979, but that was before either country became member of ASEAN.

Avoiding War Among Members

There has been little skirmishes, little tensions here and there, between, say, Myanmar and Thailand. From time to time, Thailand and Laos had their difficulties. And every time Singapore and Malaysia had a rift or diplomatic shouting match, people thought, "Oh, they are preparing for war", but there was no evidence that they were seriously contemplating war.

War has, in a sense, become unthinkable between ASEAN members. That is the true hallmark of a security community, which ASEAN is striving to become. But the real definition of a security community is a group of states that has ruled out the use of force as a means of problem-solving, which has developed long-term expectations for peaceful change in its inter-relationship. And ASEAN certainly qualifies as one of the few examples — the only other example in the developing world would be in South America — and that is a singular achievement for ASEAN.

Engaging Outside Powers

Third, ASEAN has been exceptionally successful in engaging the so-called outside powers. Now "outside" or "inside" is difficult to define, but ASEAN's real genius is to try to formalise ways whereby it can engage, rather than exclude, all outside powers and all the major powers of the international system. It has done this either within the ASEAN Post Ministerial Conference framework or ASEAN Regional Forum (ARF) framework, or simply through its diplomatic processes and initiatives.

Now this is very, very rare. Take the comparable institutions in the Middle East, for example, the Gulf States' grouping of the Gulf Cooperation Council (GCC). Now, there was a time when people compared the Gulf Cooperation Council with ASEAN, when ASEAN was facing Vietnam, and the GCC was facing Iran and Iraq, who were sort of major regional hegemons, looking down upon the weak and relatively smaller states of the Council. ASEAN similarly was facing Vietnam, which had intervened in Cambodia. But what happened? The Cold War ended, and ASEAN very quickly took in Vietnam, Cambodia, Laos and Myanmar.

So Vietnam is a part of the community today, whereas Iran and Iraq have never become part of the Gulf Cooperation Council and never will be. Now, that is engagement.

Also, ASEAN is the only organisation in the world today which engages all the major powers of the international system, whether you look at the United States, Japan, China, India or the European Union. It is the only regional organisation that has all the global actors as its dialogue partners, as members of the ARF. So ASEAN, in a sense, provides, if not structural power leadership, at least diplomatic leadership, entrepreneurial leadership through this, and it is very successful in that.

Providing a Platform for Further Institutional Building in Asia Pacific

Finally, ASEAN has done well in providing a platform for further institutional building in the Asia Pacific. Now the Asia Pacific or larger Asia as a region has never had a diplomatic regional organisation. Attempts to create other regional organisations had failed. In 1947, India tried to create the Asian Relations Organisation. There was also the Asia-Africa Conference in Bandung in 1955. China very consciously tried to create an Asian organisation. Both did not succeed.

But today, thanks to ASEAN, we have a whole range of regional organisations in the Asia Pacific. We had the Asia-Pacific Economic Cooperation (APEC) in 1989 and onwards, the ARF in 1994, and now, the East Asia Summit which will lead to a future East Asian Community, and of course, the Asia-Europe Forum. And all these institutions have used ASEAN as the platform.

Now it may be a misnomer to say that ASEAN is the leader of these institutions, but ASEAN is certainly the hub, the platform. Without ASEAN, these institutions could not have been created so easily. It was much easier for everybody to get together and create these institutions because there was already a process in ASEAN's Post Ministerial Conference. ASEAN had various diplomatic consultative mechanisms already in place and they could latch onto it.

In the 1980s or 1970s, there was not a single region-wide institution for political and economical cooperation in the Asia Pacific. Today, we have several — in fact, an alphabet soup — of such institutions, and credit goes to ASEAN.

PROBLEMS, FAILURES AND LIMITATIONS

But what about ASEAN's problems, failures or limitations? Now, the first thing that comes to mind is the unused mechanisms for conflict resolution. ASEAN is not short of institutions. There is the ASEAN High Council provided under the Treaty of Amity and Cooperation, and the ASEAN Troika. But these institutions remain unused, especially when it comes to resolving conflict. When there are conflicts, they are either settled bilaterally, or through the mediation and arbitration of the International Court of Justice and the like. And ASEAN itself has not been able to help, for better or for worse. There are good reasons for it, but these reasons are becoming increasingly difficult to sustain.

Next is ASEAN's inability to deal with internal problems of members with regional implications. There are exceptions to this, but generally ASEAN has been very reluctant to deal with or get engaged in political change or internal developments, even those whose internal developments have regional implications. Its handling of Myanmar is a good example. This needs to change.

But still, it is the deference to sovereignty, to non-interference that has made ASEAN reluctant to get involved in this. However, the fact is that the vast majority of problems that the ASEAN region faces today or the world faces today are not inter-state conflicts but internal conflicts. They are intra-state problems. And not being able to cope with these makes it difficult for ASEAN to claim the credibility of a successful regional organisation.

ASEAN also has problems dealing with transnational challenges. There are a bunch of challenges, like the Asian economic crisis of 1997, the Sars crisis of 2003, the haze that struck us in this region earlier and late last year, and of course, the problem of terrorism and the tsunami. These are problems which are neither strictly domestic nor strictly international. They are transnational.

And these problems require a collective action by a group of countries to overcome, in ways that go beyond the principle of sovereignty. Here ASEAN has a mixed record. It did wonderfully well in dealing with Sars. It did wonderfully well with the tsunami by at least providing a platform for international relief. It did less well in dealing with the 1997 economic crisis. And it did very badly in dealing with the haze last year. And it is really the haze coming year after year, decade after decade, which does not do ASEAN any credit.

Another problem is that of limited institutionalised legalisation. This is a problem that is being addressed as we speak. The ASEAN Charter is partly to address the problem. ASEAN has functioned mainly as an informal club, with its way of informalism, of avoiding strict legalistic institutions and binding decisions. So, as a result, ASEAN has limited institutional capacity and legalistic mechanisms to deal with the problems that it faces.

And finally, ASEAN also suffers from a lack of meaningful engagement with regional civil society. Now again, people talk about the *peoples'*

ASEAN, or the ASEAN People's Assembly, but the extent of this engagement, bringing in the NGOs, is not strong. I know for the ASEAN Charter, the Eminent Persons Group engaged with and listened to a lot of NGOs. So it may be changing. But generally, this is a very state-centric diplomatic institution where the common people, or the representatives of the people — the NGOs — have very little say.

FUTURE OF ASEAN

What is the future of ASEAN then? Many views have been offered on this. Some people have called ASEAN a "sunset organisation". Some have called it irrelevant. Some have said that ASEAN must re-invent itself. There is a very interesting quotation from a Vietnamese friend of ours who came to a conference the Institute of Defence and Strategic Studies (IDSS) organised recently, and he said, "ASEAN risks becoming an economic community without a single market, a political and security community without collective action, and a socio-cultural community without a common identity". Beautifully put, whether true or not.

Hence, there is a credibility problem that ASEAN needs to overcome. Now what ASEAN needs to do is to maintain a balance between state sovereignty on the one hand and integration or community-building on the other. Of course, nobody can expect ASEAN to give up sovereignty completely. At the same time, nobody expects ASEAN to maintain itself purely as a 17th-century Westphalian organisation, where no collective action takes place. So there has to be a balance being struck. And the principles, in terms of the balance that needs to be struck, and the mechanisms that will have to be in place for the balance to be maintained, is the challenge that ASEAN Charter makers face today.

Balancing or Bandwagoning?

Then there is the problem in maintaining equal footing between balancing versus bandwagoning in dealing with external powers like China, India or the United States. By balancing, I mean that ASEAN collectively counters the rise of an external power, in this case, maybe China, India or Japan. It then creates an alliance against a rising power. Bandwagoning means joining a camp, siding with either China or India against other powers.

Now ASEAN has to strike a balance between balancing and bandwagoning. However, I see this is the least of ASEAN's problems. Because, as I said, ASEAN has been very successful in engaging *all*. Binding the great powers into a common regional home has been ASEAN's singular success. That has to be maintained, for if ASEAN or its members decide that they want to break ranks and join China, or India or the United States in this balance-of-power geopolitics, it will certainly do a lot of harm to itself.

And finally, the balance between ASEAN's way of informality, of avoiding legalistic mechanisms on the one hand and legalisation and institutionalisation on the other. Now the "ASEAN way" has become a little discredited because it has not created adequate institutions to deal with the dangers ASEAN faces.

But the ASEAN way also has advantages. It creates a comfort level that is valuable, especially when you are bringing in new members and a new generation of members. So the golf games, the karaoke singing, the holding of hands that you see in ASEAN meetings are useful. But that now has to be balanced with actual legal and institutional mechanisms that are binding on states. And we will see how the ASEAN Charter performs on this.

What ASEAN Needs to Do

Now my wish list of what ASEAN could do, and here, I am speaking as a private citizen. I think ASEAN should have a charter which offers very specific and usable mechanisms or instruments for crisis management.

Usable is very important, because as I said, ASEAN is not short of mechanisms, but they have not been used. There will need to be a spelling out of situations in which the sovereignty principle can be relaxed, and more authority given to the Troika and ASEAN's Secretary-General for preventive diplomacy. ASEAN also needs to create more avenues for defence and military cooperation, including peacekeeping, which by the way, may be happening. There was an ASEAN defence ministers' meeting last year when a proposal for an ASEAN peacekeeping force was rejected. So there is more work to be done there.

And engaging civil society is absolutely crucial, both formally and informally. Beyond that, I think ASEAN also has a sort of institutional overdose problem. There are too many institutions coming up, and nobody knows what is the division of labour among them. So, it is very important for ASEAN to use new institutions such as the East Asia Summit to complement rather than duplicate or compete with existing institutions like the ARF and APEC. And we should revitalise the ARF and APEC, because they do have very important functions.

I also think it is time to really think *beyond* ASEAN to the larger region of Asia and develop a charter on governance covering, for instance, sovereignty and collective action, democracy and human rights, crisis management and human security. Such charters are coming up in Latin America, for example, the Inter-American Democratic Charter. And in Africa and the European continent, there are similar initiatives that may be good for ASEAN to look at.

For further elaboration of my wish list, I have written a paper with Professor Jorge Dominguez. We have provided a wish list of what ASEAN should do.[1]

I have also made a presentation before the Eminent Persons Group on drafting the ASEAN Charter and as Professor Tommy Koh will be centrally involved in drafting the actual charter, I pleaded with him to take into account the things that we have put forward. One of them is that ASEAN should have a rule that foreign ministers or relevant ministers must meet within 72 hours of a crisis. ASEAN should also talk about modalities for handling unconstitutional changes of government, like in a military coup. There has to be some penalty, or some kind of mechanism to create a process for the restoration of constitutional rule, such as giving more power to the ASEAN Troika and the Secretary-General.

ENDNOTE

1. Available online at http://www.ips.org.sg/events/p2007/index.htm.

CHAPTER 10

A New ASEAN

ASEAN SECRETARIAT[1]

INTRODUCTION

ASEAN is almost 40 years old. It is one of the most successful regional organisations, and is deepening its integration efforts on all fronts. However, it has reached a critical milestone in its evolution and development. The challenges facing ASEAN today are many, namely, a more complex and dynamic international environment, stiffer economic competition, greater regional interdependence, and the need to narrow the development gap among its Member Countries. There is no guarantee that it will continue to be relevant in the coming decades and remain the driving force in regional cooperation. ASEAN must address all these challenges.

It is believed that an ASEAN Charter will enable ASEAN to better position itself to overcome these challenges. The Charter presents an opportunity for ASEAN to take stock of its achievements and shortcomings, reaffirm its relevance, and forge a new path for its integration. Besides conferring a legal personality on ASEAN, the Charter will also seek to infuse ASEAN with a renewed sense of purpose, to reaffirm and codify key objectives and key principles, to strengthen its institutions and organisational structure, and strive to narrow the development gap, so that ASEAN can retain its role as a driving force in regional dialogue and cooperation.

Towards this end, on 12 December 2005, ASEAN leaders signed the Kuala Lumpur Declaration on the Establishment of the ASEAN Charter at their Summit in Kuala Lumpur. They set up an Eminent Persons Group (EPG) of senior ASEAN statesmen to examine ASEAN and come up with recommendations for an ASEAN Charter that is forward-looking and progressive. The EPG has come up with its recommendation for an ASEAN Charter and the report will be considered by ASEAN leaders at their Summit in Cebu on 13–14 January 2007.

TOWARDS A NEW ASEAN

Realising ASEAN's Vision

The EPG has reacted positively to the Member Countries' current efforts to accelerate the realisation of the ASEAN Community by 2015. However, accelerating to 2015 means that ASEAN cooperation will expand to many areas that will require changes in the way ASEAN works. It will require the strong political will of ASEAN leaders, and active support of the ASEAN people.

Stating the Objectives and Principles

As ASEAN has evolved dramatically beyond what was envisaged in the ASEAN Declaration of 1967, with the scope of cooperation now covering the political-security, economic and finance, and socio-cultural fields, the founding objectives of ASEAN will need to be updated and brought in line with the new realities confronting the Association. The EPG has been talking about how the Charter's objectives should incorporate the vision of a single market with free movement of goods, ideas and skilled talent, along with efforts to harmonise regional economic policies and strengthen regional linkages and connectivity.

In the thinking of the EPG, ASEAN's objectives should also include the strengthening of democratic values, ensuring good governance, upholding the rule of law, developing the promotion of human rights and achieving sustainable development.

ASEAN common principles, as enshrined in ASEAN's fundamental documents, such as the ASEAN Declaration (1967), the Declaration of Southeast Asia as the Zone of Peace, Freedom and Neutrality (ZOPFAN)(1971), the Treaty of Amity and Cooperation (TAC)(1976), the Bali Concord (1976), the Treaty on the Southeast Asia Nuclear Weapons Free Zone (SEANWFZ)(1995), the ASEAN Vision 2020 (1997), and the

Bali Concord II (2003), have served ASEAN well up to this day. These principles have safeguarded ASEAN's common interests and formed the foundation upon which Member Countries have developed mutual trust and *modus vivendi* which have been accepted by all. ASEAN's principles are universally recognised, and are found in the Charter of the United Nations and other basic international treaties, conventions, concords and agreements subscribed to by ASEAN Member Countries. Other principles of ASEAN relate to the Bandung Conference of 1955, and to the unique circumstances of ASEAN's founding and the importance of building trust and cooperation among Member Countries. These principles have been integral to the success of ASEAN, and they will continue to do so.

The EPG is contemplating how to improve decision-making in ASEAN. This has, hitherto, been based on consensus. The EPG is of the view that consensus decision-making should not be allowed to hold up decisions or create an impasse in ASEAN cooperation. Economic cooperation is one area where a more flexible approach can be adopted. The EPG leans towards a wider use of the "ASEAN minus X" formula. Some EPG members talked about a vote when absolutely necessary.

Raising Resources and Narrowing the Development Gap among ASEAN Member Countries

ASEAN's efforts in building the ASEAN Community will require considerable resources. It will be necessary to review ASEAN's budget to see how best it can support the new demands, taking into account the resource constraints. The principle of equal contribution is likely to be retained for ASEAN's operational expenditure, in line with the equal treatment accorded to all Member Countries. Efforts should also be made to attract more resources from the business sector, international organisations and ASEAN's partners and friends.

The development gap within ASEAN has to be addressed as it could otherwise adversely affect ASEAN's ability to achieve its goals. In this regard, the EPG has contemplated setting up a special fund to help narrow the development gap and support other ASEAN regional development. Given the implications of such a far-reaching idea, it would have to be further studied by financial and fiscal agencies.

Strengthening ASEAN's Organisational Structure

During the first ten years of its establishment, ASEAN operated without a central secretariat. Even after the ASEAN Secretariat was established in Jakarta in 1976, Member Countries were initially reluctant to create a strong central secretariat. Subsequently, this cautious approach resulted in a slow and rather piecemeal development of the whole ASEAN structure. The current institutional framework is not sufficiently well-structured to deal with the increasing number of transnational and trans-sectoral issues. ASEAN lacks effective coordination among its various bodies. The key challenge is to adopt a holistic approach to establish an overall structure that can provide unity in purpose, focus and effective implementation of ASEAN leaders' decisions, and ASEAN agreements.

The EPG is expected to propose various measures to strengthen the ASEAN Secretariat and to improve the efficiency of ASEAN meetings. One innovation is to have Permanent Representatives from ASEAN Member Countries based in Jakarta. This could settle a variety of cross-sectoral issues and reduce the need for many meetings.

Creating Legal Personality

By embarking on building the ASEAN Community, ASEAN has clearly signalled its commitment to move from an Association towards a more structured Intergovernmental Organisation, in the context of legally binding rules and agreements. In this regard, ASEAN should have a legal personality. ASEAN needs relevant privileges and immunities as are necessary for the exercise of its functions and the accomplishment of its objectives. The EPG is keen to urge all ASEAN Member Countries to put in place measures, including legislation, to give effect to such a legal personality.

Monitoring Compliance and Implementation

ASEAN must establish a culture of honouring and implementing its decision and agreements, and carrying them out on time. Delays and non-compliance will be counter-productive, undermine ASEAN's credibility and disrupt ASEAN's efforts in building the ASEAN Community. It is clear from the EPG's consultations with the incumbent Secretary-General and his two predecessors that ASEAN's problem is not one of lack of vision, ideas, and action plans. The problem is one of ensuring compliance and effective implementation of decisions. As ASEAN steps up its integration efforts, appropriate monitoring and compliance mechanisms should be established.

Promoting ASEAN as a "People-Centred Organisation"

ASEAN remains a diverse grouping of ten nations with different socio-cultural identities, norms, and varied historical experiences. But this diversity is also its strength. ASEAN leaders have recognised the importance of rallying the people of ASEAN behind ASEAN's goals. More needs to be done to promote greater awareness of ASEAN among the people, particularly through media and communications programmes. While it may be difficult to include this idea in the Charter, ASEAN leaders, the Secretary-General of ASEAN, the ASEAN Secretariat, and the ASEAN Foundation should consciously and continuously pursue this objective.

The EPG feels strongly the need for ASEAN to engage representatives of civil society, think-tanks and the ASEAN Inter-Parliamentary Assembly (AIPA, previously known as AIPO), among others, to better communicate the objectives and activities of ASEAN to the public, and to provide feedback on their current concerns. They can also be encouraged to participate in ASEAN activities and programmes revolving around the commemoration of key ASEAN activities to promote greater regional identity and consciousness, such as the ASEAN Day celebration; activities in culture, sports, arts and heritage; museum exchanges; exhibitions; publications; student and youth exchanges, and women's programmes.

Strengthening External Relations

ASEAN has, over the years, developed useful linkages with countries beyond the region through its dialogues and forums, such as the Dialogue Partnerships, the ASEAN Regional Forum (ARF), ASEAN Plus Three, and the East Asian Summit (EAS). Each of these arrangements brings unique strengths to the relationships and must therefore be nurtured. ASEAN can build on such links to ensure that it remains outward-looking and successfully pursues friendly relations and mutually beneficial cooperation with partners and friends. This will help forge a regional architecture that is open and inclusive, as well as strengthen regional cooperation to deal with the growing number of transboundary challenges ranging from transboundary haze pollution, terrorism, transnational crime, and maritime security to natural disasters and communicable diseases.

All major powers are ASEAN's Dialogue Partners, and they engage with ASEAN on regional and international affairs of common concern as well as contribute to development activities within ASEAN. The EPG believes the Charter should provide for Dialogue Partners to appoint their ambassadors to be accredited to ASEAN and based in Jakarta to facilitate and develop their relationships with ASEAN. ASEAN should also seek to retain its centrality and strengthen its role as the driving force in regional cooperation. To do so, the Secretary-General of ASEAN can be given a larger mandate to represent ASEAN's interests and to devote greater attention to nurturing cooperation with Dialogue Partners, and other regional and international organisations.

Furthermore, ASEAN must maintain close cooperation with the United Nations, where it has recently obtained Observer status. The role of the ASEAN Chair in regional processes, such as the ARF, should be enhanced to preserve ASEAN as the primary driving force. In addition, ASEAN should also strengthen the coordination and support role of the ARF Unit within the ASEAN Secretariat.

CONCLUSION

The challenges facing ASEAN in the next decade and beyond are daunting. It is clear from the EPG's discussions that ASEAN cannot allow itself to be overtaken by events. To remain relevant, ASEAN must strengthen itself to actively and effectively address and overcome the challenges. The report of the EPG will seek to strike a balance between preserving ASEAN's fundamentals and putting in place a stronger basis for ASEAN's cooperation and future integration. The ASEAN Charter should provide a framework for a stronger ASEAN. If political will is mustered, ASEAN will be reinvigorated.

ENDNOTE

1. Presented on behalf of the ASEAN Secretariat by Dr. Azmi Mat Akhir.

CHAPTER 11

The ASEAN Charter: Milestone or Illusion?

RODOLFO SEVERINO

INTRODUCTION

In talking about the ASEAN Charter, I wish to ask a few questions. The first is: why does ASEAN after 40 years need a charter? The most basic reason is that we need to give ASEAN a legal personality. That is the formal reason for it, but we also need the charter to make clear what ASEAN's objectives are, because in the last 40 years, we have been kind of muddling through.

Another question that has to be answered is this: where is ASEAN heading? Is it going for a customs union, a single market, a common market? Hopefully, the charter will also embody the principles and norms which ASEAN stands for. The charter should make clear what ASEAN really stands for. It would spell out the procedures and the decision-making processes and make them precise. At the same time, the charter would more precisely define ASEAN's institutions and their relationship to one another. The final question is: Would the charter also define the relationship between the state and its people?

THE CHARTER AND THE EMINENT PERSONS GROUP REPORT

I have not seen the report containing the views of the Eminent Persons Group (EPG) on the ASEAN Charter. Tomorrow, the Group will submit its report to the ASEAN leaders in Cebu. The document has been awaited with much anticipation since the EPG was appointed in December 2005. What is being submitted tomorrow will not be a draft charter but the recommendations of the EPG as to what the charter should contain. I do not think that I am betraying any confidences since a lot of the charter has already been leaked to the press. It seems that what the EPG report would contain would be the recommendations on the principles and objectives that ASEAN would stand for. It would speak about democratic values, good governance, the rejection of unconstitutional and undemocratic changes of government, the rule of law, human rights, and fundamental freedoms. It would also prescribe close economic cooperation and integration, aiming for a single market and the narrowing of the development gap between ASEAN members. One of the things I have read is that it mentions a "calibration" of the non-intervention policy, although I am not quite clear as to what "calibration" means. It also aims to promote a sense of common identity in Southeast Asia that would eventually evolve into an ASEAN Community and, ultimately, an ASEAN Union.

The EPG would also recommend certain changes to the structure of ASEAN, including the conversion of the ASEAN Summit into a council of leaders that would meet formally at least twice a year to give the leaders a hands-on role in the management of the Association. It also would recommend three ministerial councils that would be the top layer below the council to handle political/security, economic and social-cultural matters.

The EPG would have certain recommendations for the mobilisation of resources to support the operations of ASEAN, but it would preserve the principle of equal contribution from the member-states. Not many people know this, but in ASEAN, every single member, whether it is Singapore or Laos, whether it is Brunei or Cambodia, contributes exactly the same amount to the operation of the Association. But the EPG would also encourage voluntary contributions with the aim of narrowing the development gap between member-states.

The EPG would recommend measures to foster compliance by member-states with their obligations in ASEAN, including dispute-settlement mechanisms that would cover not only the economic agreements but all other agreements of ASEAN. Included also are enforcement mechanisms, the monitoring of compliance by states with obligatory reporting processes and, not least, redress or sanctions for non-compliance, including the suspension of the rights and privileges of membership.

To promote the effectiveness of the organisation, the EPG would recommend the increase in the number of Deputy Secretary-Generals from two to four and the appointment of Permanent Representatives of the ASEAN states to the ASEAN Secretariat in Jakarta. It would also recommend the professionalisation of staff, the improvement in the efficiency of the meetings, the invigoration of the ASEAN Foundation and the creation of an ASEAN Institute.

Clearly, the core of the recommendations is the promotion of peer pressure for compliance with the norms and commitments that the member-states have adopted and undertaken, the possibility of sanctions, and the fact that with the ASEAN Charter, the commitments of the member-states would be legally binding upon them. I think that these are all good recommendations and it is now up to the task force, whose members have all been appointed to draft the actual charter, hopefully within the year.

RED FLAGS FOR CHARTER TASK FORCE

I intend to raise some red flags on the charter, not to throw cold water on the recommendations but to indicate what the task force and those of us who are interested in these things, those of us who are on the outside looking in, need to watch out for.

First, I think that when the drafters of the charter lay down and formulate the norms and commitments that the charter would contain, they must ensure that these norms and commitments are clear and specific. Otherwise, there would be very little likelihood that the sanctions can be applied, because if the norms and commitments are too vague, there would be endless arguments as to whether a violation would require sanctions to be applied.

The charter is meant to be legally binding, but it would certainly suffer from the weaknesses of international law in general in the sense that ASEAN lacks the equivalent of the European Court of Justice or the International Court of Justice that would enforce compliance. Even the International Court of Justice does not have an army with which to enforce its decisions unless one considers the UN Security Council as the enforcement authority. There is nothing like that in ASEAN. By way of sanctions, there would only be the withdrawal of benefits and concessions by the other ASEAN members.

One reality is that many of the recommendations of the Eminent Persons Group are already in current ASEAN agreements, declarations and other commitments. However, many of them are not being implemented for national-interest reasons. For example, many of the commitments to push regional economic integration are not being complied with. There is nothing to prevent the Secretary-General from reporting on non-compliance, so it is a matter of cultural and political considerations that stands in the way between the commitment and its implementation. Hopefully, the charter, by making things legally binding, will promote reporting, but it requires a certain degree of political will and desire to overcome these political and national interests and cultural barriers. To cite another example, there is nothing to prevent the professionalisation of the Secretariat as called for by the EPG report, except the will of the member-states. We will have to see whether any provisions in the charter that call for the professionalisation of the Secretariat will actually be carried out.

Many of the recommendations, particularly those of an organisational nature, that are expected to be in the EPG report would require additional funding. For example, funding will be needed for the expansion of the functions of the Secretariat, the increase in the number of Deputy Secretary-Generals, the creation of an ASEAN Institute, the transfer of resources to the less advanced members, and the dispute settlement and compliance mechanisms. The task force would have to address the questions of where the funding is going to come from.

ORGANISATION

The EPG would also have recommendations on the organisation of the Association. Among the recommendations, as I mentioned earlier, is the installation of Permanent Representatives to the ASEAN Secretariat in Jakarta. The task force will have to make sure that there is some value added by this set of officials to the ASEAN process. And the tendency has to be resisted to accredit the ambassadors to Indonesia concurrently to the

ASEAN Secretariat, as many countries have accredited their ambassadors to Belgium as ambassadors also to the EU, in which case, there is really not much point to this recommendation.

Now the EPG would create councils of political/security affairs, economic affairs and social-cultural matters. The question is: Where do the finance ministers fit in here? I do not believe that the finance ministers would want to be part of the economic ministers and trade and industry ministers' forum. Also, who in the government set-ups would be on the socio-cultural council? Then, what is the role of the defence ministers, who constitute a new forum in ASEAN? Again, I do not envision defence ministers necessarily subordinating themselves to the foreign ministers.

POLITICAL WILL NEEDED

In order for the changes to work, the requisite political will would have to be there, political will that is not sufficiently evident up to now. Most of the recommendations that the EPG would make can actually be carried out even now, without the charter. The charter would make them more legally binding and, hopefully, strengthen the sense of obligation. Still, nothing can force a member-state to comply with its obligations, and I doubt very much whether member-states would be quick to impose sanctions on a fellow member. There would be a provision in the charter for voting as a last resort, as a way to go beyond consensus building. But I do not expect the issue of voting to be resolved very quickly.

These considerations are the reasons why the Eminent Persons Group strongly stresses that ASEAN must develop a culture of honouring and implementing its decisions, agreements and timelines. This is because, in the end, everything will depend on political will.

CHALLENGE TO TASK FORCE

The previous speaker, Dr Azmi Mat Akhir, talked about the ASEAN way and the need for it to be changed, to be modified, to be qualified. And I agree with him. The ASEAN way, of course, was adopted for good reasons and has served Southeast Asia and the larger region of East Asia well. The ASEAN way may no longer do under current global and regional conditions. And this is precisely the reason why a charter is being proposed.

But, in the end, people who are going to draft the charter must make sure that changing the ASEAN way will be for the better rather than being done just for its own sake.

THREE POSSIBILITIES FOR CHARTER

I see three possibilities for how the ASEAN Charter draft will turn out. One is that, because of the compromising that takes place, ASEAN comes out with a weak and wishy-washy charter. The second possibility is that the task force comes out with a great charter and the member-states comply with it. Of course, this is the best scenario. And perhaps even worse than the first possibility is that ASEAN comes out with a great charter and the member-states proceed to ignore it. These are some of the things I think we need to watch out for. But I am optimistic, and my optimism arises not least from the presence of Tommy Koh in the task force.

About the Contributors

Amitav ACHARYA is Deputy Director and Head of Research at the S Rajaratnam School of International Studies, Nanyang Technological University, Singapore, where he also holds a professorship. Prior to this appointment, he was a Fellow of the Harvard University Asia Center, and concurrently a Fellow of Harvard's John F. Kennedy School of Government at its Center for Business and Government. His previous academic appointments include: Professor of Political Science at York University, Toronto; Lecturer in Political Science at the National University of Singapore; and Fellow of the Institute of Southeast Asian Studies. He held an ASEM Chair in Regional Integration at the Asia-Europe Institute, University of Malaya, during 2003–2004 and was the inaugural Direk Jayanama Visiting Professor of Political Science at Thailand's Thammasat University in June–July 2005. Acharya's publications include 12 books (five self-authored and seven edited or co-edited), an Adelphi Paper for the International Institute for Strategic Studies, a Pacific Strategic Paper for the Institute of Southeast Asian Studies, and numerous other monographs, book chapters and articles in such journals as *Pacific Review, Journal of Peace Research, Journal of Strategic Studies, International Security, International Organization* and *Survival.* Among his academic books are *Constructing a Security Community in Southeast Asia: ASEAN and the Problem of Regional Order* (Routledge, 2001, Chinese translation published by Shanghai People's Press, 2004), and *Regionalism and Multilateralism: Essays on Cooperative Security in the*

Asia Pacific, 2nd edition (Eastern Universities Press, 2003). His earlier book, *The Quest for Identity: International Relations of Southeast Asia* (Oxford, 2000) was reviewed in the journal *Pacific Affairs* "as the best work on Southeast Asian regionalism available". He is co-editor of *Reassessing Security Cooperation in the Asia Pacific* (MIT Press, 2005). Acharya is a member of the editorial board of several journals, including *Pacific Review*, *Pacific Affairs*, *European Journal of International Relations* and *Global Governance*. He is one of the co-editors of the *Asian Security* monograph series published by Stanford University Press. He is a founding co-president of the Asian Political and International Studies Association. Acharya is also a regular writer and commentator on key developments in international public affairs. His public interest books include *The Age of Fear: Power Versus Principle in the War on Terror* (Rupa & Co and Marshall Cavendish, 2004). His international media appearances include interviews with CNN, BBC World Service, CNBC and Channel NewsAsia (Singapore). His recent op-eds have appeared in *Financial Times, International Herald Tribune, The Straits Times, Far Eastern Economic Review, Japan Times,* and *YaleGlobal Online,* covering topics such as Asian security, regional cooperation, the war on terror, the situation in Burma, and the rise of China and India. His areas of specialisation include regionalism and multilateralism, Asian regional security and international relations theory.

AZMI Mat Akhir, a Malaysian, holds a Bachelor of Agriculture (1973) and a Master of Agriculture (1976) from the Bogor Agricultural University, Indonesia. He received his Diploma of Advanced Studies in Soil Sciences (1980) and Doctor of Science at the International Training Centre for Post-graduate Soil Scientists, Geological Institute, State University of Ghent, Belgium. He started work with the ASEAN Secretariat on 1 January 1993. Before joining the ASEAN Secretariat, he served as a civil servant at the Department of Agriculture for Peninsular Malaysia from 18 October 1976 to 31 December 1992. During this period, he held various professional and supervisory positions, starting from a junior Agriculture Officer to head of sections, and finally, as Assistant Director for Agricultural Planning and Development. With the ASEAN Secretariat, he started as a Senior Officer for Trade and Commodities, moved up to Assistant Director and Head of Food, Agriculture and Forestry Unit, and then to Director of Bureau of Resources Development. Since 1 August 2005, he has been assigned as Special Assistant to the Secretary-General of ASEAN for Institutional Affairs and Special Duties. Dr Azmi is the longest-serving member of the current group of ASEAN Secretariat's openly recruited professional staff. As of today, he has served the ASEAN Secretariat for 14 years.

Manu BHASKARAN is an Adjunct Senior Research Fellow at the Institute of Policy Studies. He is also concurrently Partner and Member of the Board, Centennial Group Inc, a policy advisory group based in Washington D. C., where he heads the Group's economic research practice. Mr Bhaskaran co-leads the institute's work in the area of economics. His major area of research interest is the Singapore economy and the policy options it faces. Prior to his current positions, Mr Bhaskaran worked for 13 years at the investment banking arm of Societe Generale as its Chief Economist for Asia. He began his professional career at Singapore's Ministry of Defence, focusing on regional security and strategic issues. Mr Bhaskaran graduated from Cambridge University with a Masters of Arts and also has a Masters in Public Administration from Harvard University.

Beatrice CHIA-RICHMOND is one of Singapore's most prominent theatre directors and actresses. Her works are often a combination of strong text, provocative approach and purposeful direction. Ms Chia-Richmond graduated from the Guildhall School of Music & Drama in London with a BA (Hons) in Performing Arts. She began her professional career in the United Kingdom before returning to Singapore in 1998. In Singapore, she embarked upon a multi-faceted career in television, film and theatre. She was also the host of Singapore's longest-running weekly arts programme *Art Nation.* Ms Chia-Richmond began directing in 2001. For her critically-acclaimed directorial debut, *Shopping and F***ing,* she was awarded Best Director at the Life! Theatre Awards. Some of the theatre productions that she is proudest of are *Bent,* which won Play of the Year at the Life! Theatre Awards, and most recently, *Cabaret,* a musical staged at the Esplanade in 2006. She is currently Artistic Director of Toy Factory Productions, one of Singapore's most vibrant and productive bilingual theatre companies. She is also Singapore's Young Artist Award recipient for 2006.

CHUA Hak Bin, based in Singapore, joined Citigroup in May 2006 as a Director and covers economic developments in Asia with a special focus on equity themes. Prior to joining Citigroup, Dr Chua was the Senior Regional Economist (Southeast Asia) and Asian Equity Strategist at DBS Group. Dr Chua has previously worked at the Monetary Authority of Singapore for six years, during which time he headed the External Economies Division, Planning, Policy & Communications Division, and the Financial Surveillance Division. Dr Chua also has extensive experience in the corporate world, having been General Manager (Corporate Finance) in a plantation, property and construction company in Malaysia. Before that, Dr Chua was an Economist at RHB Research Institute in Kuala Lumpur. Dr Chua received a PhD and MA in Economics from Harvard University. He holds a BS in Engineering and BA in Economics (magna cum laude) from Brown University. He was a Visiting Lecturer and Research Fellow at the Economic Growth Centre at Yale University for a year following the completion of his PhD.

Cherian GEORGE is an Adjunct Senior Research Fellow at the Institute of Policy Studies. He is based at the Wee Kim Wee School of Communication and Information at the Nanyang Technological University, Singapore, where he is Acting Head of Journalism and Publishing. He is also a Research Associate of the Asia Research Centre at Murdoch University. Dr George's main research interest is in the press and politics, especially the alternative media. He is the author of *Contentious Journalism and the Internet: Towards Democratic Discourse in Malaysia and Singapore*, published with the support of the IPS in 2006; and *Singapore: The Air-Conditioned Nation*, published in 2000. Before taking up his appointment at Nanyang Technological University, Dr George was a post-doctoral Fellow at the Asia Research Institute, National University of Singapore. He holds a PhD in Communication from Stanford University. He is also a graduate of Cambridge University, where he read social and political sciences, and the Columbia University Graduate School of Journalism. Before moving to academia, he spent 10 years at *The Straits Times*, where he wrote mainly on domestic politics and media issues. Dr George continues to practise journalism as the editor and publisher of *What's Up*, a monthly current affairs newspaper for children.

KWOK Kian Woon holds a PhD from the University of California at Berkeley. He is Associate Professor, Associate Chair (Academic) and Head, Division of Sociology at the School of Humanities & Social Sciences, Nanyang Technological University, Singapore. His teaching and research interests relate to social and cultural transformations in the contemporary world. His current research topics include: violence, intolerance, security and risk in the early 21st century; religion, secularity, and pluralism in the post-September 11 world; social memory and civil society; comparative cultural policy (arts, heritage and creative cities); and Singapore studies. His writings in books, journals and newspapers include two chapters in *The Encyclopedia of the Chinese Overseas* and numerous essays on culture, cultural policy, and social issues in the post-September 11 world. He is actively involved in the public sector and in civil society, including as Member of the National Heritage Board and the Singapore Art Museum Board, Honorary Chairman, Board of the National Archives of Singapore, Co-Chairman of the Steering Committee for the Singapore Biennale 2006, Member of the Steering Committee on the National Art Gallery (and Chairman of its Museological Advisory Group), and the immediate past president of the Singapore Heritage Society.

LIM Sun Sun holds a PhD and an MSc (Distinction) in Media and Communications from the London School of Economics. She is an Assistant Professor at the Communications and New Media Programme, National University of Singapore. Her research interests are the social implications of new media and public perceptions of new technology. Her current research focuses on the domestication of infocomm technologies by middle-class families in Asia, media literacy amongst youths, public perceptions of new media piracy and online privacy. She is also a member of the National Internet Advisory Committee, which advises Singapore's Media Development Authority on issues relating to new media.

Rodolfo SEVERINO is a Visiting Senior Research Fellow at the Institute of Southeast Asian Studies in Singapore and a frequent speaker at international conferences in Asia and Europe. Having been Secretary-General of ASEAN from 1998 to 2002, he has completed a book, entitled *Southeast Asia in Search of an ASEAN Community* and published by the

Institute of Southeast Asian Studies, on issues facing ASEAN, including the economic, security and other challenges confronting the region. His views on ASEAN and Southeast Asia have been published in *ASEAN Today and Tomorrow*, a compilation of his speeches and other statements. He writes articles for journals and for the press. As a member of the faculty at the Asian Institute of Management in the Philippines in the school year 2003-2004, he lectured on regional economic cooperation, the elements of competitiveness, and leadership in the management of change. Before assuming the position of ASEAN Secretary-General, Severino was Undersecretary of Foreign Affairs of the Philippines. In the Philippine Foreign Service, Severino was Ambassador to Malaysia from 1989 to 1992, Chargé d'Affaires at the Philippine Embassy in Beijing from 1975 to 1978, Consul General in Houston, Texas, and an officer at the Philippine Embassy in Washington, D. C. Between overseas postings, he worked as special assistant to the Undersecretary of Foreign Affairs and Assistant Secretary for Asian and Pacific Affairs at the Department of Foreign Affairs. He twice served as ASEAN Senior Official for the Philippines. Before joining the Philippine Government, Severino worked at the United Nations and with Operation Brotherhood-Laos. He has a Bachelor of Arts degree in the humanities from the Ateneo de Manila and a Master of Arts degree in international relations from the Johns Hopkins University School of Advanced International Studies.

Brenda S. A. YEOH is Professor at the Department of Geography and Head of the Southeast Asian Studies Programme, National University of Singapore. She is also Research Leader of the Asian Migration Research Cluster at the University's Asia Research Institute. Her research interests include the politics of space in colonial and post-colonial cities; and gender, migration and transnational communities. Her first book was *Contesting Space: Power Relations and the Urban Built Environment in Colonial Singapore* (Oxford University Press, 1996; reissued Singapore University Press, 2003). She also published *Singapore: A Developmental City State* (John Wiley, 1997, with Martin Perry and Lily Kong); *Gender and Migration* (Edward Elgar, 2000, with Katie Willis); *Gender Politics in the Asia Pacific Region* (Routledge, 2002, with Peggy Teo and Shirlena Huang); *Toponymics: A Study of Singapore Street Names* (Eastern Universities Press, 2003, with Victor R. Savage);

Theorising the Southeast Asian City as Text (World Scientific, 2003, with Robbie Goh); *The Politics of Landscape in Singapore: Construction of "Nation"* (Syracuse University Press, 2003, with Lily Kong); *Approaching Transnationalisms* (Kluwer, 2003, with Michael W. Charney and Tong Chee Kiong); *State/Nation/Transnation: Perspectives on Transnationalism in the Asia Pacific* (Routledge, 2004, with Katie Willis); *Migration and Health in Asia* (Routledge, 2005, with Santosh Jatrana and Mika Toyota); *Asian Women as Transnational Domestic Workers* (Marshall Cavendish, 2005, with Shirlena Huang and Noor Abdul Rahman); and *Working and Mothering in Asia* (NUS Press and NIAS Press, 2007, with Theresa Devasahayam).